KT-230-363

DK Guide to
SPACE

DORLING KINDERSLEY
LONDON • NEW YORK • MOSCOW • SYDNEY
www.dk.com

A DORLING KINDERSLEY BOOK
www.dk.com

This book is dedicated to two young stars,
Jennifer and Elizabeth Burton

Project Editor Ben Morgan
Project Art Editor Martin Wilson
Managing Editor Mary Ling
Managing Art Editor Rachael Foster
Designer Robin Hunter
DTP Designer Almudena Díaz
Production Lisa Moss
Picture Researcher Andy Sansom

First published in Great Britain in 1999
by Dorling Kindersley Limited
80 Strand, London WC2 ORL
A Penguin Company

8 10 9 7

Copyright © 1999 Dorling Kindersley Limited, London
First paperback edition 2004

All rights reserved. No part of this publication may be reproduced,
stored in a retrieval system, or transmitted in any form or by any
means, electronic, mechanical, photocopying, recording, or otherwise,
without the prior written permission of the copyright owner.

A CIP catalogue record for this book
is available from the British Library

Paperback edition ISBN-13: 978-0-7513-3925-3
            ISBN-10: 0-7513-3925-3
Hardback edition ISBN-13: 978-0-7513-5877-3
            ISBN-10: 0-7513-5877-0

Reproduced in Italy by
G.R.B. Editrice, Verona

Printed and bound in China
by Toppan Printing Co. Ltd

See our complete catalogue at www.dk.com

CONTENTS

STARGAZERS

PEOPLE HAVE BEEN FASCINATED by the stars for thousands of years, but it was not until telescopes were invented that astronomers began to find out what the Universe is really like. Our home planet is nothing more than a tiny speck of matter floating in a gigantic void called space. The stars we see all around make space look crowded, but they are incredibly far away, and incredibly far apart. Space, on the whole, is empty. It is also big. Too big, in fact, to imagine. Even the closest star would take more than a million years to reach at the speed of Concorde. Distances in space are so vast that astronomers measure them in light years. One light year is the distance light travels in a year – 10 million million kilometers (6.2 million million miles). Thanks to modern telescopes, astronomers can now see stars and galaxies billions of light years away.

ASTRONOMY TODAY
Professional astronomers spend far more time sitting in front of computers than staring through telescopes. In a modern observatory, all the data collected by the telescope is fed into a computer. The computer then processes the data to produce images that highlight particular features. The digital images can be downloaded onto the Internet so that astronomers worldwide can see them.

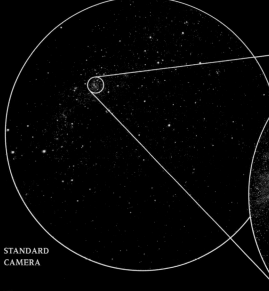

STANDARD
CAMERA

UK SCHMIDT TELESCOPE

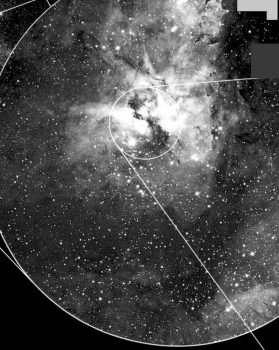

ANGLO-AUSTRALIAN TELESCOPE

ZOOMING IN
Stars and galaxies look tiny because they are so far away. The Carina Nebula is actually a gigantic dust cloud, but from Earth it looks like a tiny red blob. It is just visible to the naked eye from Earth's Southern Hemisphere. By using powerful telescopes, astronomers can zoom in on the Carina Nebula to study it in minute detail. Buried deep in its heart is an exploding star called Eta Carinae (*see also* p. 42). Light from the Carina Nebula takes 8,000 years to reach Earth, so these pictures show the Nebula as it was 8,000 years ago.

STANDARD TELESCOPE

RADIO TELESCOPE

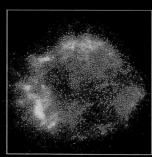

X-RAY TELESCOPE

SEEING THE INVISIBLE
Just like Superman, astronomers use X-ray vision to see things that are invisible to the human eye. The three pictures to the left show the wreckage of a star that exploded in a normal telescope shows an empty patch of sky, but radio and X-ray telescopes reveal a cloud of debris billions of kilometres across. Superhot gases in the cloud are spreading out into space at up to 6,000 km (3,700 miles) per second.

HUBBLE VISION

Earth's atmosphere distorts the light from stars, making them difficult to see clearly. The Hubble Space Telescope gets around this problem because it floats in space 595 km (370 miles) above Earth, well out of reach of the atmosphere. Hubble is about the size of a school bus and orbits Earth at 28,000 kmh (17,500 mph). It can see galaxies billions of light years away, so it sees them as they were billions of years ago.

WORKING TOGETHER

Sometimes, several telescopes are combined to give a sharper view of a faint object. The Very Large Array (above) in New Mexico, USA, is made up of 27 huge dishes that detect radio waves. A computer combines all the data to make a single image. The curved dishes work by reflecting radio waves onto a central detector. Light-detecting telescopes work the same way, but their dishes have a mirrored surface to reflect light.

HEAVENLY TWINS

The twin Keck telescopes in Hawaii are the largest light-detecting telescopes in the world. They sit on top of an extinct volcano, far away from city lights. Here, the clear, thin air is ideal for observing faint galaxies. The mirrors inside the Keck telescopes are made of separate sections that can change position. This allows astronomers to change the mirrors' shape for the best view.

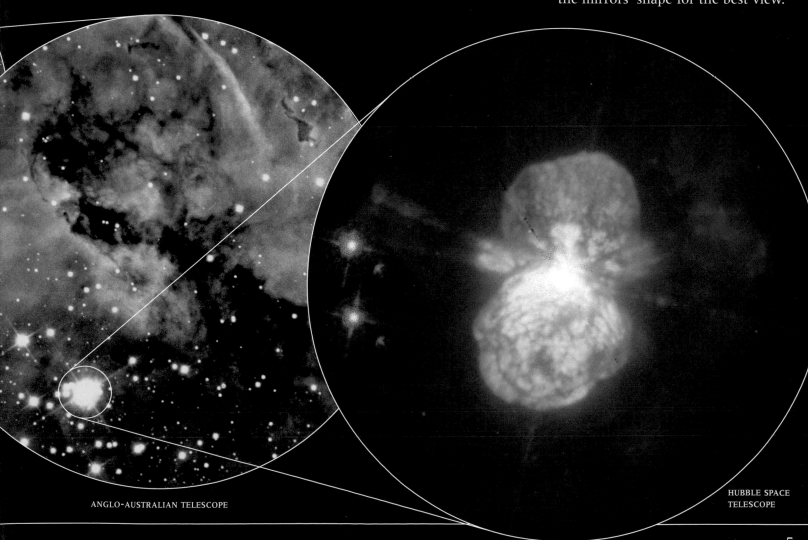

ANGLO-AUSTRALIAN TELESCOPE

HUBBLE SPACE TELESCOPE

THE SOLAR SYSTEM

The SOLAR SYSTEM IS OUR NEIGHBOURHOOD in space. At its centre is the Sun, our local star, which takes up 99.9 per cent of the Solar System's mass. The Sun's gravity keeps everything else trapped near it, including nine planets and their 63 moons. The planets race around the Sun along paths called orbits, spinning like tops as they move. The four inner planets are balls of rock and metal. The outer planets are giant balls of gas or liquid, except icy Pluto, the outermost. Pluto is 6 billion km (4 billion miles) from the Sun. The nearest star is 7,000 times further away.

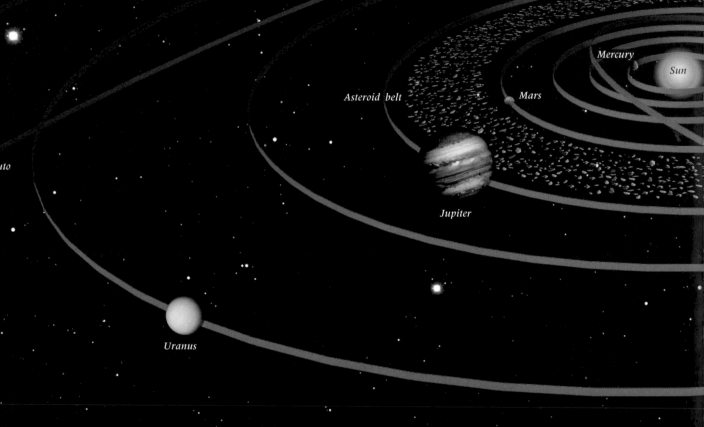

Pluto

Uranus

Asteroid belt

Mars

Mercury

Sun

Jupiter

SUN
The Sun is a huge, spinning ball of luminous gas, more than a million times larger than Earth. Its energy comes from nuclear reactions in the core, where temperatures reach 15 million °C.

MERCURY
This small, cratered planet is the closest to the Sun. During the day its surface is a scorching 430°C, but at night it falls to –170°C. A "year" on Mercury – the time it takes to orbit the Sun – lasts only 88 Earth days.

VENUS
Venus is Earth's ugly sister. Its size and structure are similar to Earth's, but the air is poisonous and the surface is far too hot to support life. Venus spins around so slowly that its "day" is longer than its "year".

EARTH
Our home planet is the only place in the Universe known to support life. Oceans of water cover two-thirds of the rocky surface and white clouds of water swirl through the atmosphere. Earth has only one moon.

MARS
People once thought there was life on Mars. Like Earth, it has mountains, canyons, icy poles, and an atmosphere. However, its surface is a barren desert. Mars is about half as wide as Earth and has two moons.

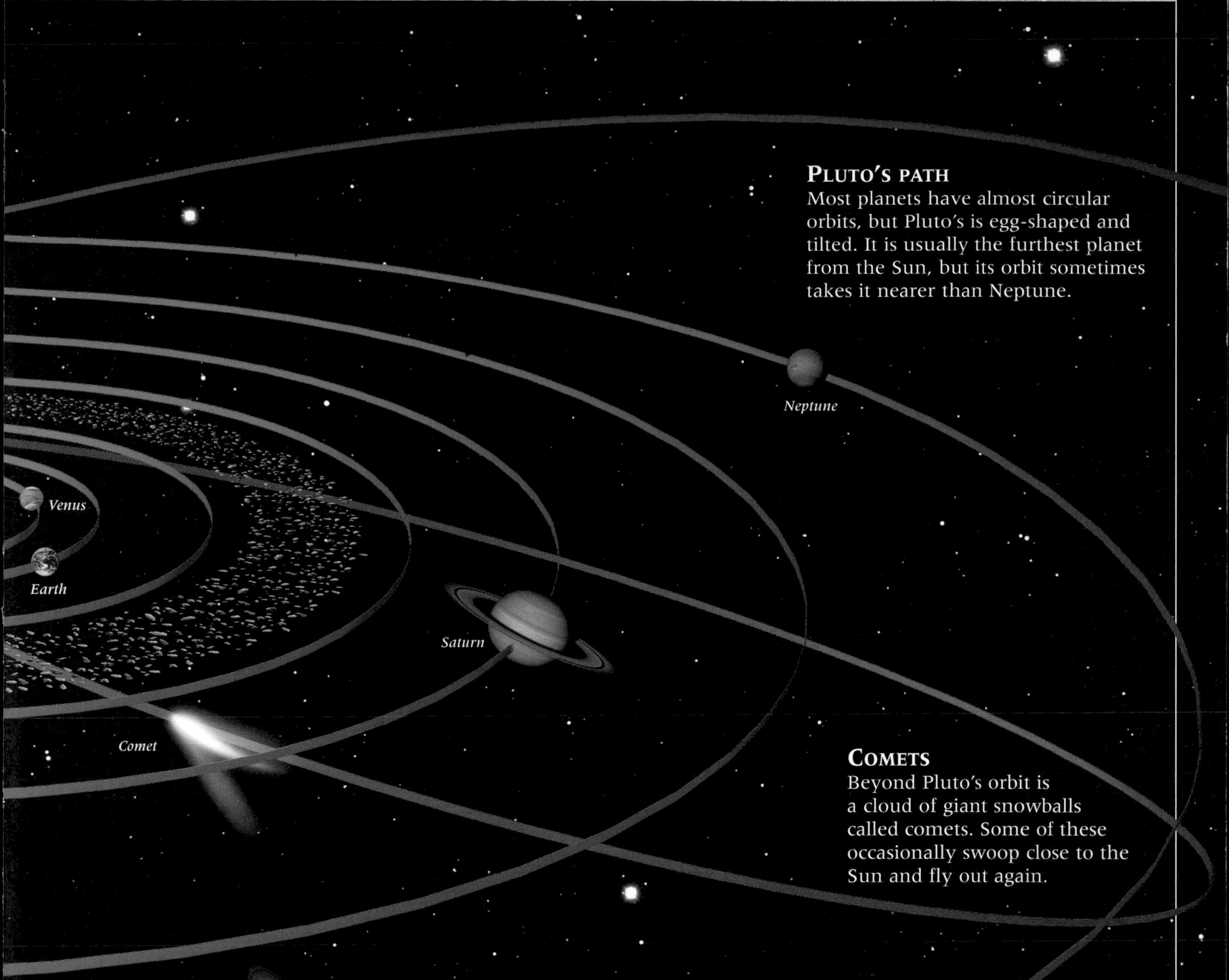

PLUTO'S PATH
Most planets have almost circular orbits, but Pluto's is egg-shaped and tilted. It is usually the furthest planet from the Sun, but its orbit sometimes takes it nearer than Neptune.

Neptune

Venus

Earth

Saturn

Comet

COMETS
Beyond Pluto's orbit is a cloud of giant snowballs called comets. Some of these occasionally swoop close to the Sun and fly out again.

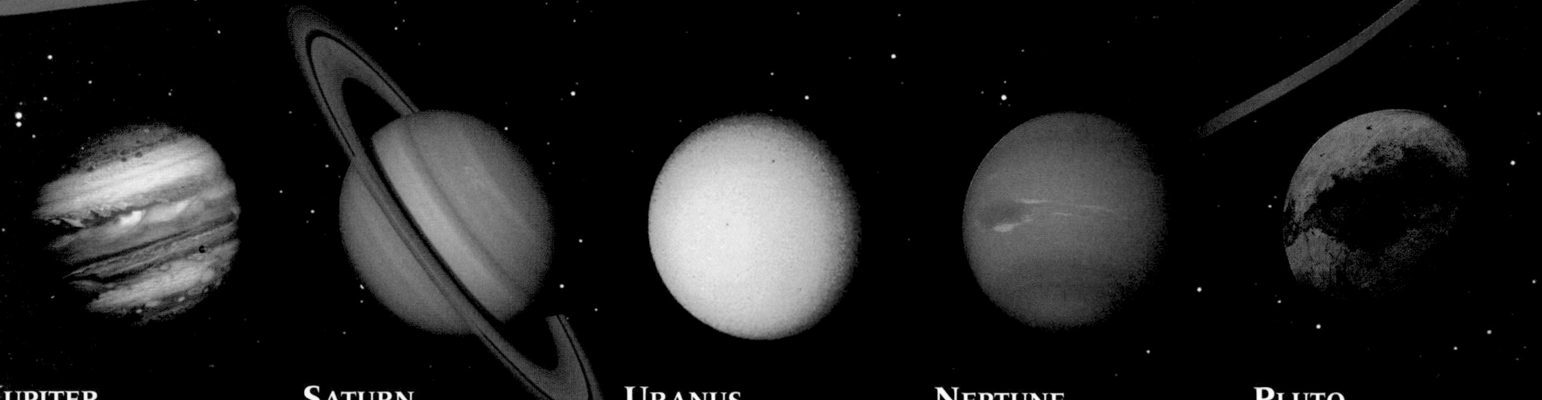

JUPITER
The largest planet, Jupiter is heavier than all the other planets put together. It is a huge globe of liquid and gas, covered by circular bands of cloud. Jupiter has 16 moons, including four bigger than Pluto.

SATURN
If you put all nine planets in a giant bucket of water, only Saturn would float. It is almost as big as Jupiter but far lighter. Its rings are made of millions of dazzling chunks of ice. Saturn also has 18 icy moons.

URANUS
With a featureless blue surface, Uranus looks like a billiard ball. It is four times wider than Earth, 20 times further from the Sun, and has 15 moons. The man who discovered Uranus wanted to call it "George's Star".

NEPTUNE
The winds on Neptune are the fastest in the Solar System, tearing around the planet at up to 2,000 kmh (1,240 mph). Raging storms appear as dark spots. Neptune is similar to Uranus but has only eight moons.

PLUTO
Mysterious Pluto is too small and far away to see clearly. It gets so little sunlight that its air freezes in winter. Its year is 248 Earth years – a person born here would not live to be a single year old. Pluto has one moon.

THE SUN

O UR SUN IS A TYPICAL STAR. Like any star, it is a gigantic ball of shining hydrogen gas. Its source of power lies buried deep in the central core, where the temperature soars to 15 million °C. A pinhead this hot would set fire to everything within 100 km (62 miles). The core is like a continually exploding nuclear bomb – but much, much bigger. Every second, it turns 4 million tonnes of hydrogen into pure heat and light. The gas around the core absorbs this energy and bubbles up to the seething surface, where the heat and light escape into space.

SPOTTY FACE

Dark spots come and go on the Sun's face all the time. Called sunspots, they are dark because they are a few thousand degrees cooler than the gas around them. By watching sunspots move, astronomers have worked out that the Sun's equator spins around much faster than its poles.

THE CORONA

Around the Sun is a wispy atmosphere called the corona. It extends millions of kilometres into space. The Sun's surface is a million times brighter than the corona, yet the corona is millions of degrees hotter. No one knows why. Gas escapes through holes in the corona and streams into space at up to 3 million kmh (2 million mph).

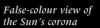

False-colour view of the Sun's corona

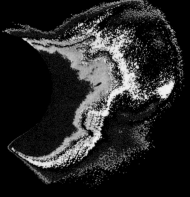

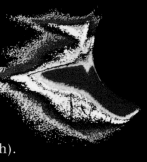

STORMS ON THE SUN

Scientists call the superhot gas in the Sun plasma. Colossal bubbles of plasma sometimes blast out of the corona during a storm. The photographs above show a bubble erupting in a matter of minutes. Occasionally, the blastwave from a sunstorm heads directly for Earth. Our atmosphere protects us from the worst effects, but even so, sunstorms can wreck satellites or overload electricity lines, causing powercuts.

ARCHES OF FIRE

Huge arches of plasma leap hundreds of thousands of kilometres from the Sun and hang suspended in space for months on end. The largest ever recorded was longer than the distance from Earth to the Moon. These arches, called solar prominences, follow the Sun's magnetic field. Sometimes they break away and launch billions of tonnes of plasma into space.

An eruption of superhot gas (plasma) escapes the Sun's massive gravity.

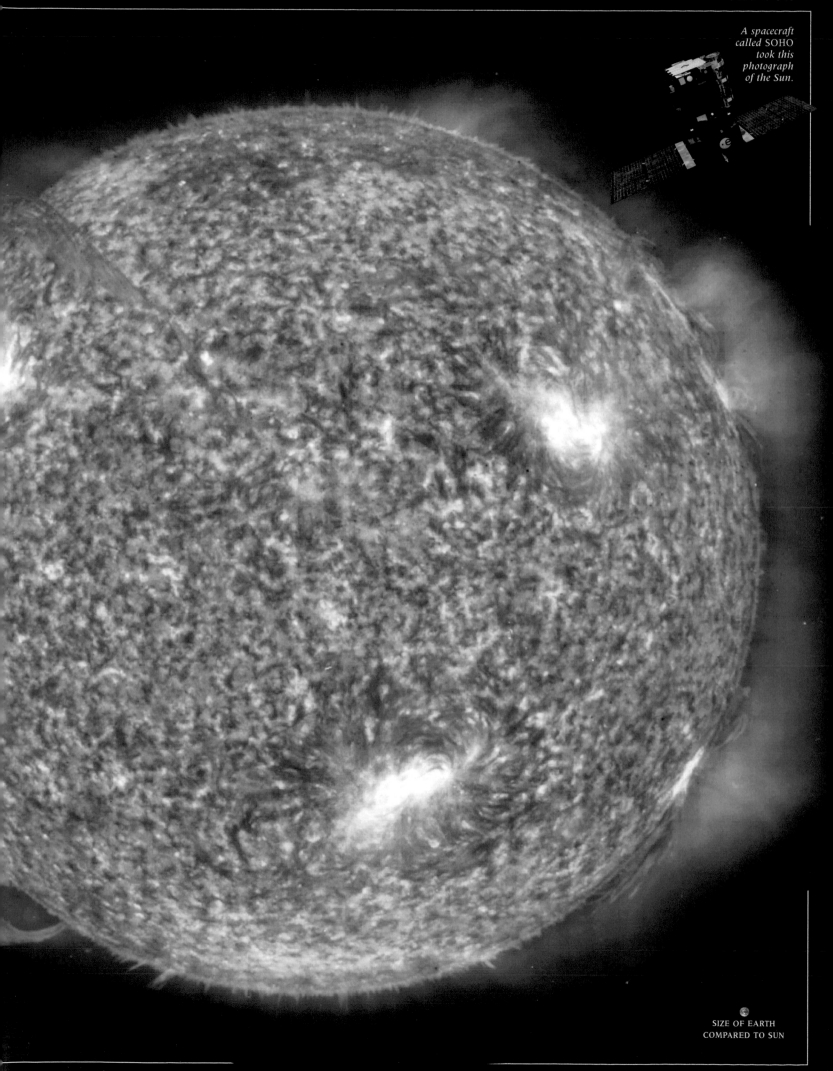

A spacecraft called SOHO took this photograph of the Sun.

SIZE OF EARTH
COMPARED TO SUN

ECLIPSES

A TOTAL ECLIPSE OF THE SUN is a spectacular sight. For a few minutes, day turns into night. The sky goes dark, stars and planets appear, streetlights come on, and birds stop singing – creating one of the most dramatic events in nature. Eclipses take place when light from the Sun or Moon is blocked for a short time. Lunar eclipses happen quite often. Eclipses also occur on other planets, for example when Jupiter's moons move into the planet's shadow. These can only be seen with a telescope. When viewing a solar eclipse, it is vital to remember that you should never look at the Sun without proper eye protection.

STAGES OF A SOLAR ECLIPSE

After first contact, it takes about an hour for the Moon to completely cover the Sun.

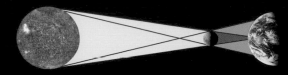

The Sun's light travels towards the Moon and Earth. (Not to scale)

The Moon moves between the Sun and Earth.

The centre of the Moon's shadow causes a total solar eclipse.

A SOLAR ECLIPSE
During a solar eclipse, the Moon moves in front of the Sun. For a few minutes the bright Sun disappears as the Moon's shadow quickly crosses the Earth's surface. Anyone inside the narrow shadow sees a total eclipse. Anyone just outside it sees a partial eclipse.

TOTALITY
The stage of a solar eclipse when the Moon completely covers the Sun (below) is called "totality". This can last for up to eight minutes. During totality, the Sun's faint outer atmosphere, the corona, looks like pearly white streamers around the dark centre.

DIAMOND RING

Just before totality, the last sunlight shines through valleys on the Moon. When this happens, a bright spot appears at the edge of the Moon, producing a stunning effect known as a diamond ring. This lasts for only a few seconds. The picture below also shows flares of hot gas (red) leaping above the Sun's surface.

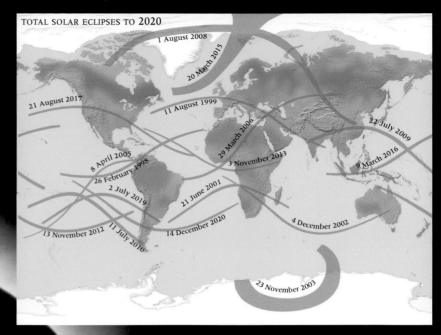

TOTAL SOLAR ECLIPSES TO 2020

1 August 2008
20 March 2015
21 August 2017
11 August 1999
22 July 2009
8 April 2005
29 March 2006
3 November 2013
9 March 2016
26 February 1998
2 July 2019
21 June 2001
13 November 2012
11 July 2010
14 December 2020
4 December 2002
23 November 2003

IN THE SHADOW OF THE MOON

There are at least two solar eclipses every year, but most of them are partial – the Moon only covers some of the Sun. Total solar eclipses occur somewhere in the world about every 18 months, but they often fall over oceans or sparsely populated areas. Although the Moon's shadow travels thousands of kilometres over the Earth during an eclipse, its path is often less than 100 km (62 miles) wide. Total eclipses only occur in the same place roughly every 330 years.

THE DARK SIDE

During a lunar eclipse the Earth's shadow may take up to four hours to cross the face of the Moon. The total eclipse may last over an hour. Some sunlight is bent by Earth's atmosphere and still reaches the Moon, making it appear red. Sometimes there are three lunar eclipses in one year. In other years there may be none. A lunar eclipse can be seen from anywhere on the night side of Earth.

During a lunar eclipse, the Moon gradually becomes much darker, and often turns a reddish colour.

The Sun's light travels towards Earth. (Not to scale)

Half of Earth is in daylight. The night side is in shadow.

The Moon moves into Earth's shadow. A lunar eclipse occurs.

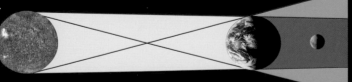

A LUNAR ECLIPSE

Sometimes the Moon passes through Earth's shadow. This happens at the time of a full Moon. Lunar eclipses do not happen every time there is a full Moon because the Moon's orbit is tilted and so it often misses Earth's shadow.

MERCURY

THE CLOSEST PLANET TO THE SUN, Mercury is scorching by day but freezing at night. If you landed on Mercury's day side, the Sun would look two and a half times larger than it does from Earth, but the sky would be black because Mercury has no air. Planet Earth would be a big, bluish "star". The Sun's gravity keeps Mercury racing around it along an oval orbit, but the planet rotates sluggishly as it moves. Mercury's face bears witness to a violent history. Countless asteroids and comets have pounded this planet, covering it with craters. There are even craters within craters within craters.

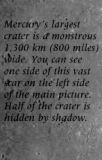

Mercury's largest crater is a monstrous 1,300 km (800 miles) wide. You can see one side of this vast scar on the left side of the main picture. Half of the crater is hidden by shadow.

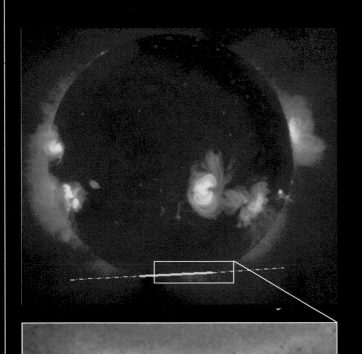

Mercury

04 49 54

CAUGHT IN THE GLARE
Mercury is always close to the Sun in the sky. This makes it very hard to see with the naked eye because the Sun's glare hides it. Astronomers can see Mercury more clearly by using powerful telescopes. The pictures above were taken with an X-ray telescope, which detects X-rays instead of normal light. They show Mercury as a tiny dark spot passing in front of the Sun's enormous face.

MARINER PROBE
The only space probe that has been to Mercury is *Mariner 10*. It arrived in 1974 and flew past the planet three times before its fuel ran out. Unfortunately, *Mariner 10* saw the same face during each flyby, so large parts of Mercury remain a mystery.

SIZE COMPARED
TO EARTH

BIRTHDAY PLANET

Mercury takes only 88 Earth days to orbit the Sun, giving it the shortest year in the Solar System. Its day (from sunrise to sunrise) lasts twice as long, so if you were born on Mercury you would get two years older every day. The long day, combined with Mercury's oval orbit, has another strange effect. On some parts of the planet you would see the Sun rise at dawn, dip below the horizon, then rise again.

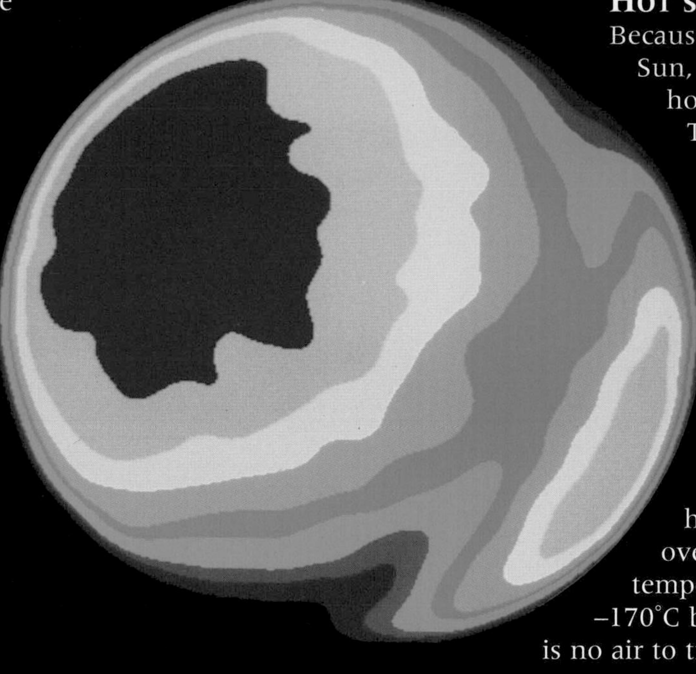

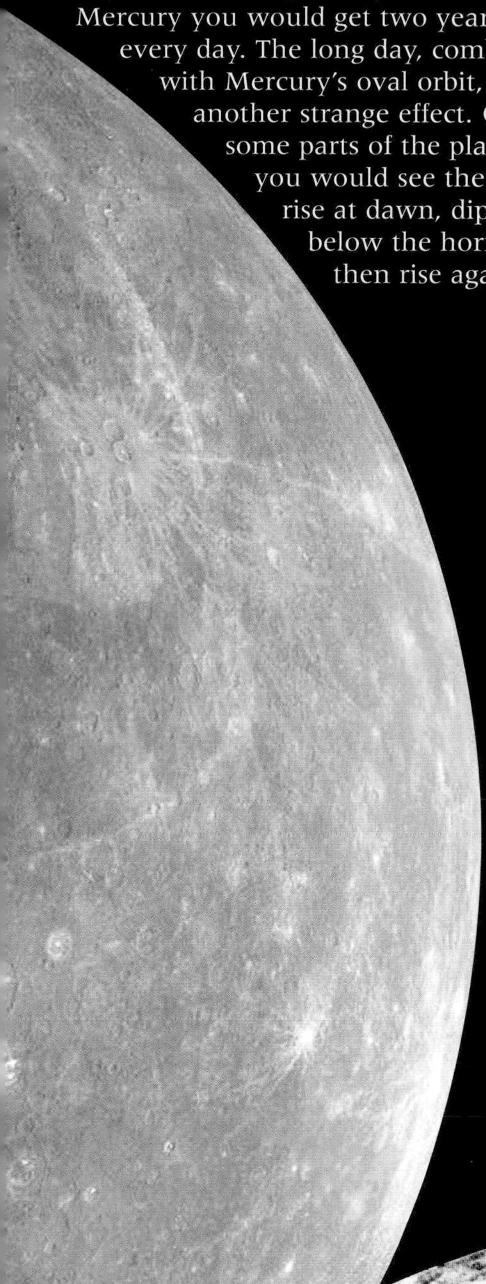

Most of Mercury's features are named after famous people in the arts, for example the Mozart crater in the Beethoven region.

Some craters on Mercury have hills in the middle. These formed when the ground "bounced back" after an asteroid collision.

HOT SPOTS

Because it is close to the Sun, Mercury gets much hotter than Earth. This heat image shows the surface temperature on Mercury. The hottest place (red) is the region around the equator in daytime. Temperatures here can reach a scorching 430°C – hotter than an oven. At night the temperature falls to –170°C because there is no air to trap heat.

LUNAR LOOKALIKE

Close up, Mercury looks just like our Moon – airless and pockmarked with craters. In some places, tall cliffs run across the surface for hundreds of kilometres, as shown in the picture below. These cliffs formed when the young planet cooled and shrank, making the surface wrinkle. Mercury's interior is a gigantic ball of iron that takes up 80 per cent of the planet's mass. This vast iron core makes it one of the densest planets.

Cliffs

VENUS

ASTRONOMERS SOMETIMES call Venus the nearest thing to hell in our Solar System. Clouds of acid cover the surface, where the temperature is high enough to melt lead. The air is so dense that it would crush a human in seconds. Venus spins the opposite way to Earth, so the Sun rises in the west and sets in the east. The planet turns very slowly, taking 243 Earth days to rotate once. This makes its "day" longer than its "year" (225 Earth days).

*This photograph was taken by the **Pioneer** spacecraft, which orbited Venus in 1979.*

SWIRLING CLOUDS

The clouds on Venus are so thick that direct sunlight cannot reach the planet's surface. The light they reflect makes Venus the second brightest object in our night sky after the Moon. The clouds are made of concentrated sulphuric acid. Even if astronauts could survive being crushed and roasted on Venus, they would soon die from acid burns. Raging winds blow the clouds right around the planet every four days, creating dark swirls in the cloud tops.

LAVA LAND

Scientists used radar data to reconstruct this panorama on a computer. The central peak is Maat Mons, the tallest volcano on Venus. It towers over a lifeless desert of solidified lava. Like most features on Venus, Maat Mons is named after a woman – in this case, the Egyptian goddess of truth and justice. No one knows if the volcano is active.

Deep circular valleys have formed where the crust has pulled apart.

Blue areas are flat plains covered by vast lava flows.

SIZE COMPARED TO EARTH

The giant volcano Maat Mons lies in the middle of the highlands.

BENEATH THE CLOUDS

This false-colour radar map shows the surface of Venus with the clouds stripped away. Solidified lava covers about 85 per cent of the planet. The blue areas show where lava floods have spread out to make vast, lowland plains. If Venus had water, these would be oceans. The white highlands, equivalent to Earth's continents, are a jumble of mountains, volcanoes, and ridges, cut by deep valleys.

TOUCHDOWN ON VENUS

In March 1982, a Russian spacecraft called *Venera 13* parachuted onto Venus and photographed its surface. The pictures revealed an orange desert littered with broken rocks – probably the remains of an old lava flow. The light level was like a cloudy day on Earth. This picture also shows an ejected camera cover (left), a colour chart, and the metal landing platform. The craft survived only 127 minutes before Venus's lethal atmosphere destroyed it.

PANCAKE DOMES

There are all sorts of strange volcanic structures on Venus, such as these seven overlapping domes of rock. Each is about 25 km (16 miles) wide and 750 m (2,500 ft) tall. Called pancake domes, they are probably small volcanoes made from thick, sticky lava that solidified before it could flow far. As the lava cooled, the domes shrank and their tops cracked.

The volcano Maat Mons is about 8 km (5 miles) tall.

The surface temperature on Venus is a scorching 482°C.

EARTH

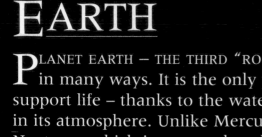

PLANET EARTH – THE THIRD "ROCK" FROM THE SUN – is unique in many ways. It is the only planet scientists know that can support life – thanks to the water in its oceans and the oxygen in its atmosphere. Unlike Mercury, which is intensely hot, and Neptune, which is extremely cold, Earth sits at an ideal distance from the Sun. For its size, Earth is very heavy due to the large iron core at its centre. The churning motion of the liquid iron creates a powerful magnetic field that shields Earth from harmful particles streaming out of the Sun. Earth's atmosphere also screens out dangerous radiation from the Sun.

RESTLESS EARTH
The Earth is not a solid ball, but is made up of many different layers. On the surface is a thin shell of solid rock – the crust. This forms the continents and ocean floors. Heat carried upwards from the central core forces sections of the crust, called plates, to move. As this happens many of the Earth's natural features are created or changed.

This image shows the Earth's crust with the water drained away.

EARTH'S ATMOSPHERE
Water covers about 70 per cent of the Earth's surface. It keeps temperatures moderate and releases water vapour into the atmosphere. The mixture of water vapour and other gases wrapped around the Earth creates the atmosphere. Its swirling white clouds are constantly moving, blown round the planet by the wind.

From space, the continents appear dark green or brown, and the oceans appear blue. As most of the Middle East is free of clouds, the deserts of North Africa and the Arabian peninsula are visible.

A WATERY PLANET

Water on the Earth takes many different forms. It is found as a liquid (in rain, lakes, rivers, and oceans), as a gas (invisible water vapour), and solid (as ice). Water not only makes life on Earth possible, but also shapes the land. This photo shows a network of channels as the River Ganges enters the Bay of Bengal. The flat land has been built by river sediment deposited over thousands of years.

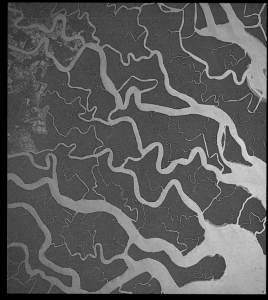

Ganges Delta seen from the Space Shuttle

MOUNTAIN RANGES

Great mountain ranges generally occur where two of Earth's plates have collided and one is forced up to form high peaks. Mt Everest in the Himalayas, seen here from the Space Shuttle, is the highest peak above sea level. However, Mauna Loa, a volcanic island in Hawaii, is Earth's tallest mountain, measured from its base on the ocean floor to its highest point.

Wildlife in Yosemite National Park, USA

Mt Everest lies on the border of China/Nepal.

EARTHQUAKES AND VOLCANOES

The boundaries of Earth's plates can be dangerous places to live. Major earthquakes occur where plates collide or slide past each other. In cities such as San Francisco, large-scale earthquakes have produced more energy than an atomic bomb. Volcanoes erupt when molten rock escapes to the surface, exploding lava and ash over huge areas. Here Klyuchevskaya volcano in Russia belches a cloud of brown ash.

TEEMING WITH LIFE

Life exists almost everywhere on Earth – from the highest mountains to the deepest underground caves. No one knows how many species of living things there are on Earth, but the total runs into millions. The first primitive forms of life probably appeared in the oceans about 3.8 billion years ago. Animals did not evolve until about 600 million years ago, while modern humans appeared only 100,000 years ago.

This view of Earth was taken from the Apollo 17 spacecraft on its journey to the Moon.

An erupting volcano, seen from the Space Shuttle

THE MOON

OUR NEAREST NEIGHBOUR in space – the Moon – has been orbiting the Earth for about 4.5 billion years. One theory suggests that the Moon was born when a Mars-sized object smashed into the young Earth and scattered a huge cloud of material into space. This eventually came together to form the Moon. Because the Moon is smaller than Earth it has less gravity and cannot "hold on" to any air or water. With no atmosphere to protect its surface, all objects hurtling towards it make a deep impact. The dark areas on the Moon are called maria, or "seas", and were formed when meteorites hit with such force that lava welled up and solidified on the crater floors.

EARTHRISE
The *Apollo 8* astronauts were the first people to see Earth rise and set as they orbited the Moon in December 1968. They described the blue and white Earth as an oasis in space compared with the grey, barren surface of the Moon. The oceans can be clearly seen here.

FULL MOON

WANING GIBBOUS

WAXING GIBBOUS

PHASES OF THE MOON
The Moon does not produce light of its own. We only see it because the Moon acts like a huge mirror, reflecting light from the Sun. As the Moon orbits the Earth it seems to change shape because we see different amounts of its sunlit side. The different shapes are called phases and the Moon takes 29.5 days to pass from one full Moon to the next.

LAST QUARTER

FIRST QUARTER

WANING CRESCENT

WAXING CRESCENT

NEW MOON
(NOT VISIBLE FROM EARTH)

SEA OF SERENITY

SEA OF TRANQUILLITY

Landing site of Apollo 11, 20 July 1969

SEA OF NECTAR

CRATERS
The surface of the Moon is covered with craters that vary in size from small pits to giant basins more than 200 km (124 miles) across. These craters are generally circular, with high walls made of debris thrown out during impact. Many have central peaks where the crust rebounded after the collision. Young craters have "rays" of ejected material that spray out as bright streaks.

EARTH

In this diagram, the size of Earth and the Moon, and the distance between them, are shown to scale.

The distance from Earth to the Moon is 384,400 km (239,000 miles)

RESOURCES FROM THE MOON

As well as metals – such as titanium, iron, and magnesium – the Moon contains a rare gas called helium-3 that may, one day, solve Earth's energy shortage. This false-colour image, taken from the *Galileo* spacecraft, shows the titanium-rich soils of the Sea of Tranquillity in blue. Most of the lunar highlands appear orange/red, indicating a low titanium and iron content.

LUNAR PROSPECTOR

In 1998, the *Lunar Prospector* spacecraft went into orbit around the Moon. It discovered vast deposits of ice near the lunar poles – probably from comets that had crashed into the Moon. The ice remains frozen because the floors of deep craters in the polar regions are always in cold shadow.

This young crater on the far side (named Giordano Bruno) is surrounded by rays that stretch 400 km (248 miles) from its centre.

SEA OF CRISES

Lunar highlands occur between maria.

SEA OF FERTILITY

LOMONOSOV

SEVER

MARE CRISIUM

FLEMING

PASTEUR

TSIOLKOVSKY

THE FAR SIDE

As the Moon spins exactly once during each orbit of the Earth, we always see the same side. The far side remained a mystery until the Russian *Luna 3* space probe travelled behind the Moon and sent back the first pictures in 1959. It looks very different from the near side – with more craters but fewer dark seas. This is because the crust is thicker and able to withstand greater bombardment. Many craters are named after scientists, such as Fleming.

This view of the Moon was taken from Apollo 11 *and mainly shows the side we see from Earth. Towards the right is a glimpse of the far side.*

MOON

Travelling at speeds of up to 38,916 kmh (24,182 mph) Apollo 11 *took three days to reach the Moon.*

At 62,550 km (38,870 miles) from the Moon, spacecraft escape Earth's gravity and are pulled towards the Moon.

MAN ON THE MOON

IN 1961, PRESIDENT JOHN F. KENNEDY announced that the United States of America would place a man on the Moon by the end of the decade. On 20 July 1969, that promise was fulfilled. More than 500 million people watched TV in astonishment as astronaut Neil Armstrong became the first person to set foot on the Moon. *Apollo 11* – and a further five missions – placed a total of 12 men on the Moon between 1969 and 1972. The dramatic, near-fatal explosion during *Apollo 13*'s outward journey made it the only cancelled trip. During a total of 80 hours outside the landing craft, the astronauts collected rock samples, took photographs, and set up experiments to find out more about the Moon's surface and interior.

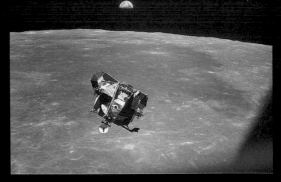

With no wind or rain to erase them, footprints made by astronauts will remain in the Moon's dust for millions of years.

LUNAR MODULE

"The *Eagle* has landed!" reported the astronauts, as the lunar module (LM) first touched down. As well as landing astronauts on the Moon, the LM was their home there. After three days on the Moon, the *Eagle* took Buzz Aldrin and Neil Armstrong back to the orbiting command module (CM). Its landing legs were left behind when it blasted off to dock with the CM for the journey home.

Apollo 16 astronauts explored the lunar highlands in their Moon buggy.

JUMPING ON THE MOON

Living in an environment with very little gravity can be a strange experience. On Earth, an astronaut in a spacesuit would weigh 135 kg (300 lb) – on the Moon, he weighs only 23 kg (50 lb). In this picture, John Young salutes the Stars and Stripes while effortlessly jumping several feet off the ground. Without any wind, the flag had to be stiffened with a rod to make it hang properly. Behind the flag, you the lunar module and the Moon buggy can be seen.

FOUR WHEELS AND OFF ROAD

On their last three trips, the astronauts took a Lunar Roving Vehicle, or Moon buggy. This battery-powered car, which folded up into a storage bay for the journey, allowed them to explore up to 26 km (16 miles) from the lunar module. The buggy had solid tyres to cope with the ultimate off-road terrain, and a top speed of almost 18 kmh (11 mph). At the rear, there was a tool kit and bags for carrying samples.

The dish antenna relayed TV pictures back to Earth.

The lunar rover was steered with a bar-shaped hand control instead of a wheel.

Television camera

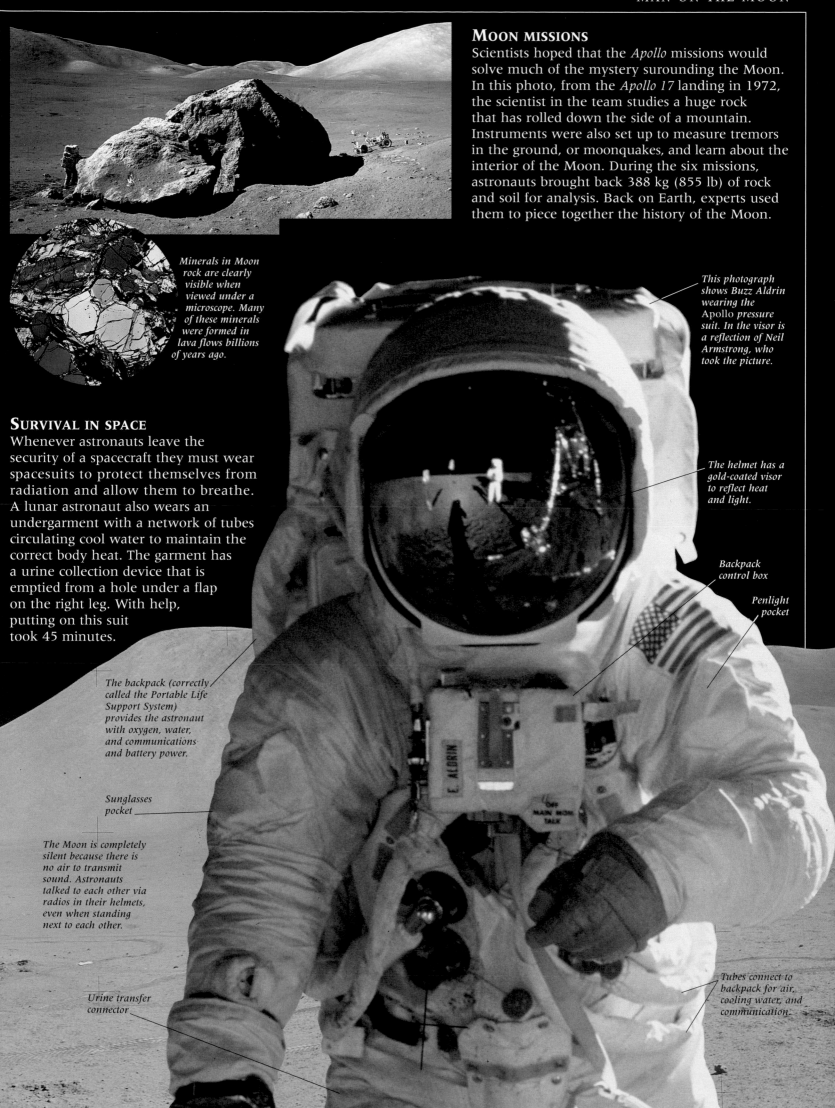

MOON MISSIONS

Scientists hoped that the *Apollo* missions would solve much of the mystery surounding the Moon. In this photo, from the *Apollo 17* landing in 1972, the scientist in the team studies a huge rock that has rolled down the side of a mountain. Instruments were also set up to measure tremors in the ground, or moonquakes, and learn about the interior of the Moon. During the six missions, astronauts brought back 388 kg (855 lb) of rock and soil for analysis. Back on Earth, experts used them to piece together the history of the Moon.

Minerals in Moon rock are clearly visible when viewed under a microscope. Many of these minerals were formed in lava flows billions of years ago.

This photograph shows Buzz Aldrin wearing the Apollo pressure suit. In the visor is a reflection of Neil Armstrong, who took the picture.

SURVIVAL IN SPACE

Whenever astronauts leave the security of a spacecraft they must wear spacesuits to protect themselves from radiation and allow them to breathe. A lunar astronaut also wears an undergarment with a network of tubes circulating cool water to maintain the correct body heat. The garment has a urine collection device that is emptied from a hole under a flap on the right leg. With help, putting on this suit took 45 minutes.

The helmet has a gold-coated visor to reflect heat and light.

Backpack control box

Penlight pocket

The backpack (correctly called the Portable Life Support System) provides the astronaut with oxygen, water, and communications and battery power.

Sunglasses pocket

The Moon is completely silent because there is no air to transmit sound. Astronauts talked to each other via radios in their helmets, even when standing next to each other.

Urine transfer connector

Tubes connect to backpack for air, cooling water, and communication.

MARS

OUR NEIGHBOURING PLANET in the Solar System has fascinated astronomers for hundreds of years. In some ways, it is very similar to Earth. It has a 25-hour day, volcanoes, valleys, and polar icecaps that expand and shrink with the seasons. Billions of years ago, Mars also had an atmosphere that could have sustained life. Dry, winding valleys etched into its surface indicate that rivers once flowed there. Over time, however, the atmosphere gradually escaped into space, and the planet grew cold. Today, Mars is a freezing desert.

The vast canyon in the middle of this picture is long enough to stretch from Paris to Athens.

These dark circles are massive volcanoes. Their craters are visible from space.

The hazy white clouds over Mars are made of frozen carbon dioxide and water ice.

THE RED PLANET

You can easily see Mars as a bright red "star" on clear nights. The colour reminded the ancient Romans of blood spilled in battle, so they named the planet after their god of war, Mars. In fact, Mars gets its colour from iron oxide – rust – in the rocks and dust that cover the surface. More recently, astronomers mistook the darker parts of Mars for vegetation because they changed shape with the seasons. In fact, these areas are just bare rock and windblown dust.

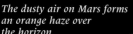

The dusty air on Mars forms an orange haze over the horizon.

RED SKY ALL DAY

Martian air is 100 times thinner than Earth's air and contains fine dust that makes the sky orange-red. Violent storms sometimes whip up huge clouds of dust from the ground and hide the planet's surface from view. The wind speed during these storms can reach 400 kmh (249 mph). The air on Mars is mainly carbon dioxide, which is poisonous – visitors from Earth would have to wear spacesuits to breathe.

SIZE COMPARED TO EARTH

MARTIAN MOONS

Mars has two tiny moons called Phobos and Deimos. They may be asteroids that became trapped by the planet's gravity. Phobos is only 28 km (17 miles) across at its widest point. Its gravity is so weak that an astronaut on Phobos would be 1,000 times lighter than on Earth and would have difficulty staying on the ground. Deimos is 14 km (9 miles) wide – the smallest known moon in the Solar System.

DEIMOS

PHOBOS

FROSTY POLES

Mars may have been warm long ago, but today it is freezing, with an average temperature of –63˚C. This picture shows a cap of frozen carbon dioxide – "dry ice" – covering the South Pole in summer. It is about 400 km (249 miles) wide. In winter the icecap expands as air freezes onto it. Mars's North Pole is covered by dry ice and water ice.

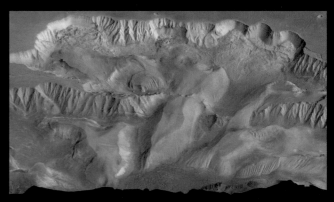

COLOSSAL CANYON

The dark slash across the face of Mars is a canyon four times deeper and ten times longer than Earth's Grand Canyon. The picture above shows a tiny part of it. Towering cliffs up to 6 km (4 miles) high surround the valley floor. In some places, the cliffs have collapsed into giant landslides. The canyon is called Valles Marineris. It was produced by slow movements in the planet's crust that split the surface open over millions of years.

MOUNT OLYMPUS

Mars is home to the most colossal volcano known to science – Olympus Mons (Mount Olympus). It is wider than England and three times taller than Mount Everest. Olympus Mons stopped erupting billions of years ago. Despite its impressive size, it was probably not a terrifying sight when it was active. The lava seeped out over thousands of years, gradually building up to form the record-breaking volcano.

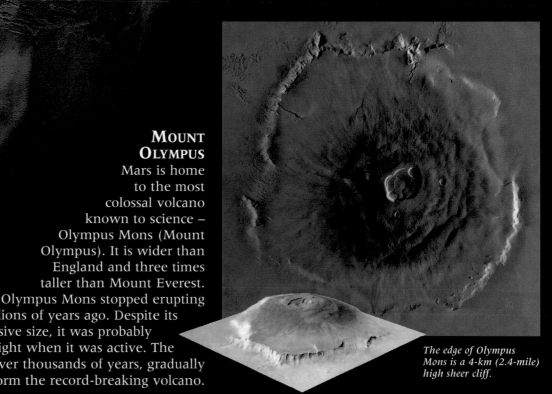

The edge of Olympus Mons is a 4-km (2.4-mile) high sheer cliff.

EXPLORING MARS

ASTRONOMERS HAVE LONG DREAMED of sending people to Mars, but so far only robotic probes have landed there. The first mission to the planet's surface took place in 1976, when the two *Viking* landers made a successful touchdown. More recently, the *Pathfinder* probe made a dramatic landing in a bouncing airbag. Thanks to such missions, scientists now know more about the "red planet" than any other planet apart from Earth. But one mystery remains: was there ever life on Mars?

VIKING INVADERS
After their year-long flights from Earth, the *Viking* landers parachuted onto two carefully selected locations on Mars. They beamed back the first-ever pictures of the surface, revealing a dusty, rock-strewn landscape and an orange sky. The landers took about 4,500 photographs of Mars's surface and studied how the planet's weather changed with the seasons. *Viking 1* continued to operate for more than six years.

VIKING LANDER'S VIEW OF A MARTIAN LANDSCAPE

Rotating camera

Satellite dish

Weather monitor

Robotic arm and scoop

LOOKING FOR LIFE
The two *Viking* landers looked for signs of life on Mars. Each lander had a robotic arm to scoop up soil. An on-board laboratory then analysed the soil for the telltale organic chemicals that living organisms produce. The results were disappointing – neither lander found any evidence that life has ever existed on Mars.

VIKING LANDER BEING TESTED ON EARTH

TWIN PEAKS
Pathfinder landed in 1997, tantalizingly close to two hills, nicknamed the Twin Peaks. The hills were only 1 km (0.6 miles) away, yet they were too far for *Pathfinder*'s rover vehicle to reach. They proved to be an important landmark, enabling scientists to pinpoint *Pathfinder*'s exact location. The taller of the two is 46 m (150 ft) high.

MARTIAN PANORAMA
The *Pathfinder* probe used a rotating camera to take the picture below. This is the view you would see if you stood on Mars and turned around in a full circle. Rocks and red dust cover everything as far as the eye can see. Scientists think that the rocks in this area were dumped by a massive flood millions of years ago.

Martian sand dunes   *Aerial*   *Solar panels*   *Hills called Twin Peaks*   *Ramp for* Sojourner

Sojourner was about the size of a microwave oven. It drove only a few metres each day.

Aerial

Large solar panel for power

Rock analyser

Suspension

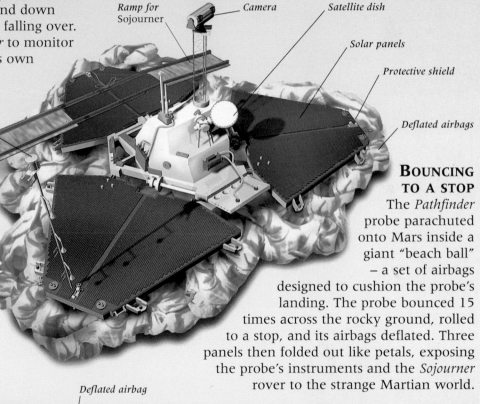

BLUE SUNSET

The sky on Mars looks orange-red during the day due to dust in the atmosphere. However, at sunset the sky in the area just around the Sun turns blue. Because there is so much dust high into the atmosphere, the sky glows for more than an hour after sunset.

MARS BUGGY

The *Pathfinder* probe carried a remote-controlled buggy called the *Sojourner* rover. *Sojourner* explored the area around the landing site, using on-board instruments to analyse chemicals in Martian rocks. Its six wheels could move up and down separately so it could trundle over rocks without falling over. Scientists on Earth used the camera on *Pathfinder* to monitor where *Sojourner* went, but the rover could use its own cameras to find its way around independently.

Ramp for Sojourner

Camera

Satellite dish

Solar panels

Protective shield

Deflated airbags

These hills may have been shaped by floods millions of years ago.

Weather monitor

BOUNCING TO A STOP

The *Pathfinder* probe parachuted onto Mars inside a giant "beach ball" – a set of airbags designed to cushion the probe's landing. The probe bounced 15 times across the rocky ground, rolled to a stop, and its airbags deflated. Three panels then folded out like petals, exposing the probe's instruments and the *Sojourner* rover to the strange Martian world.

Rover tracks

Sojourner *rover*

Deflated airbag

JUPITER

JUPITER IS BY FAR THE LARGEST PLANET in the Solar System – more than 1,300 Earths could fit inside it. Scientists call Jupiter a "gas giant" because it is mostly hydrogen. The surface layers of hydrogen gas are spread out by Jupiter's fast rotation into bands of constantly swirling cloud – it takes the planet only ten hours to make one complete turn. Below the cloud there is no solid surface. The gas simply merges into a vast ocean of hot, liquid hydrogen. Deeper still, the hydrogen is like liquid metal. Scientists think the planet's core is a ball of molten rock six times hotter than the surface of the Sun.

SUICIDE PROBE

In December 1995, the *Galileo* spacecraft dropped a small probe into Jupiter. The probe slammed into the cloud tops at 170,000 kmh (106,000 mph). Within two minutes, it had slowed down enough for a parachute to open. For the next hour it analysed chemicals in the clouds and sent radio signals back to the orbiting *Galileo* spacecraft. As the probe sank deeper, the hot atmosphere melted and crushed it.

Swirling winds reach 400 kmh (250 mph) in the Great Red Spot.

GREAT RED SPOT

Jupiter's most famous feature is the Great Red Spot, a colossal hurricane that has been blowing nonstop for at least 300 years. At 25,000 km (16,000 miles) across, it is more than twice the size of Earth. Turbulent winds circle around the hurricane, making ripples in the clouds. The Spot turns anticlockwise, taking about 12 days to rotate. Its colour probably comes from small amounts of the chemical phosphorus.

SIZE OF EARTH COMPARED
TO JUPITER AND ITS MOONS

CALLISTO

GANYMEDE   EUROPA

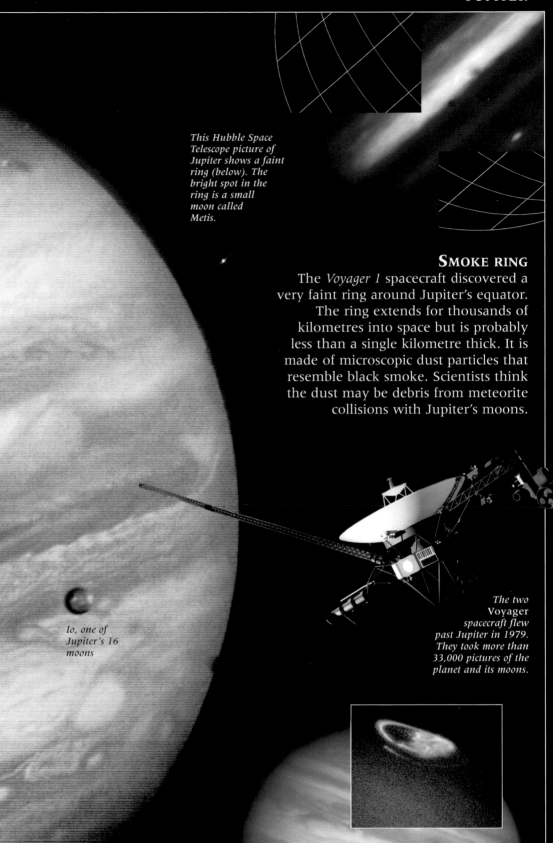

This Hubble Space Telescope picture of Jupiter shows a faint ring (below). The bright spot in the ring is a small moon called Metis.

SMOKE RING

The *Voyager 1* spacecraft discovered a very faint ring around Jupiter's equator. The ring extends for thousands of kilometres into space but is probably less than a single kilometre thick. It is made of microscopic dust particles that resemble black smoke. Scientists think the dust may be debris from meteorite collisions with Jupiter's moons.

The two Voyager spacecraft flew past Jupiter in 1979. They took more than 33,000 pictures of the planet and its moons.

Io, one of Jupiter's 16 moons

GLOWING POLES

If you could stand on Jupiter's cloud tops at night you would see a remarkable light show. Lightning bolts 10,000 times more powerful than any on Earth flash through the clouds, and brillant auroras light up the poles. The auroras are caused by charged particles spewed out of volcanoes on Io, one of Jupiter's 16 moons. The particles are sucked into Jupiter's poles by the planet's intense magnetic field.

Jupiter's auroras, photographed in ultraviolet light by the Hubble Space Telescope, appear as glowing ovals near the poles.

JUPITER'S MOONS

JUPITER'S MASSIVE GRAVITY holds in its power a set of 16 moons. Some are bigger than planets, but astronomers call them moons because they orbit Jupiter instead of orbiting the Sun. You can easily see the four largest moons if you look at Jupiter with binoculars on a clear night. Their names are Ganymede, Callisto, Io, and Europa. Jupiter's other 12 moons are more like asteroids than planets. They are small, boulder-shaped, and pockmarked with craters. The largest of the small moons, Amalthea, is only 270 km (168 miles) across at its widest point. The baby of the family is Leda, just 16 km (10 miles) wide.

SCARFACE
Callisto is the most battered world in the Solar System. Thousands of meteorites blasted this moon's surface billions of years ago, leaving craters everywhere. Since then, it has barely changed. The bright area near the top of the picture is Valhalla, the largest crater. The meteorite that made this scar must have been colossal because its shockwave left circular ripples stretching up to 3,000 km (1,900 miles) across.

KING OF THE MOONS
Ganymede is the largest moon in the Solar System, bigger even than the planets Pluto and Mercury. It is a mixture of rock and ice, with white craters on its surface where meteorites have smashed into the icy interior. The dark zones on Ganymede are ancient, cratered landscapes that have not changed in billions of years. The paler zones are smooth and have probably been filled in by flowing ice.

The Galileo spacecraft has been orbiting Jupiter and studying the planet's moons since 1995.

Callisto is a frozen mixture of ice and rock. Its temperature varies from −118°C in the day to −193°C at night.

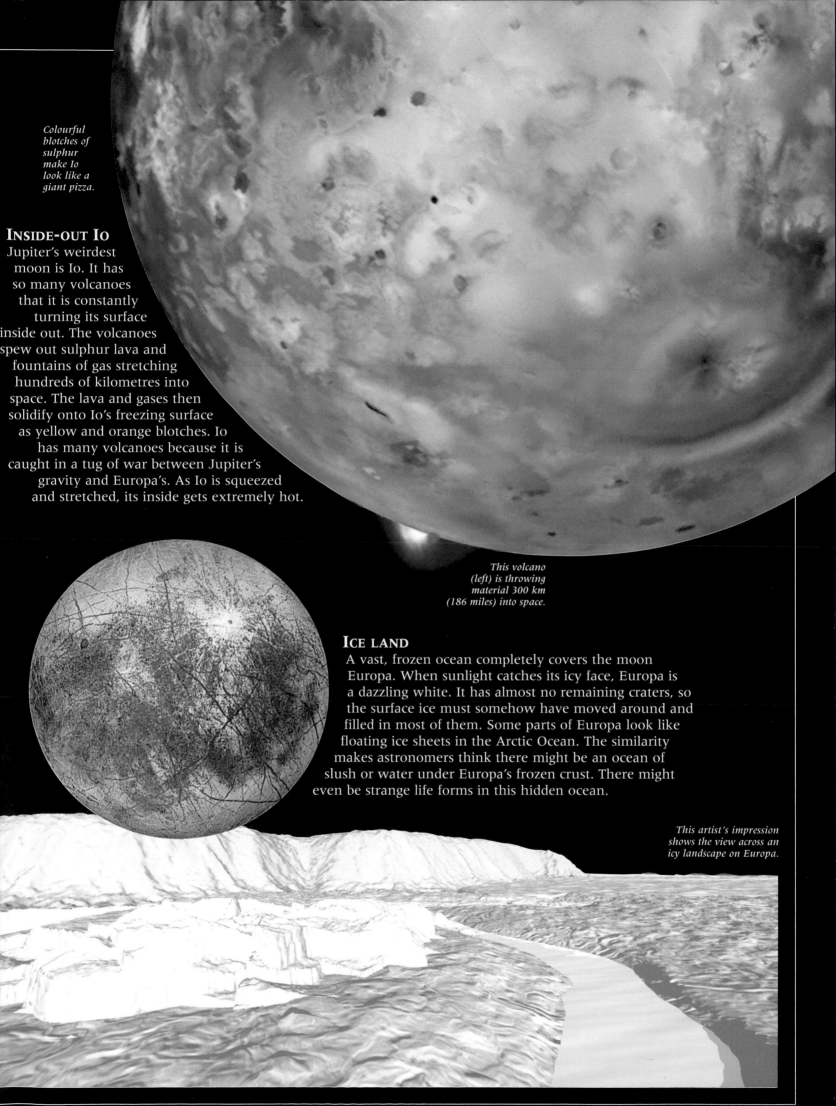

Colourful blotches of sulphur make Io look like a giant pizza.

INSIDE-OUT IO

Jupiter's weirdest moon is Io. It has so many volcanoes that it is constantly turning its surface inside out. The volcanoes spew out sulphur lava and fountains of gas stretching hundreds of kilometres into space. The lava and gases then solidify onto Io's freezing surface as yellow and orange blotches. Io has many volcanoes because it is caught in a tug of war between Jupiter's gravity and Europa's. As Io is squeezed and stretched, its inside gets extremely hot.

This volcano (left) is throwing material 300 km (186 miles) into space.

ICE LAND

A vast, frozen ocean completely covers the moon Europa. When sunlight catches its icy face, Europa is a dazzling white. It has almost no remaining craters, so the surface ice must somehow have moved around and filled in most of them. Some parts of Europa look like floating ice sheets in the Arctic Ocean. The similarity makes astronomers think there might be an ocean of slush or water under Europa's frozen crust. There might even be strange life forms in this hidden ocean.

This artist's impression shows the view across an icy landscape on Europa.

SATURN

SATURN IS BEST known for having the largest and most spectacular rings, but it holds several other records. It spins so fast that it bulges at its equator, making it the flattest planet. It also has the most moons. Astronomers have named 18 so far, but there are probably several more. The rings themselves may be debris from an icy moon that shattered when a comet smashed into it. Saturn is also the least dense planet – if you could put all the planets in a bucket of water, Saturn is the only one that would float.

SIZE COMPARED TO EARTH

POLAR LIGHTS

Saturn has a powerful magnetic field that forms an invisible bubble around the planet. Normally, the bubble shields Saturn from the solar wind – a stream of electrically charged particles that rush out of the Sun at up to 3 million kmh (2 million mph). However, some of the particles in the solar wind get sucked into Saturn's magnetic poles. When they smash into the upper atmosphere, they produce rings of light called auroras.

Saturn dwarfs its moons. Here you can see the moons Tethys (above) and Dione (below right) as pale specks. The black dot on Saturn is the shadow of Tethys.

A RINGSIDE VIEW

Saturn's rings consist of millions of sparkling lumps of ice, ranging from the size of a sand grain to the size of a house. The rings are so wide and bright that you can see them from Earth with a small telescope. No one knows how Saturn got its rings. Perhaps they are debris from a moon that was smashed by a comet, or the remains of a moon that was torn apart by Saturn's gravity. This false-colour picture shows that the rings are divided into hundreds of tiny ringlets. The gaps have been swept clean by the gravity of Saturn's moons.

Saturn is a giant spinning blob of liquid and gas. It is the second largest planet after Jupiter.

TOUCHDOWN ON TITAN

In 2004, a space probe will parachute down through the dense orange clouds that cover Titan, Saturn's largest moon. Titan is the only moon in the Solar System that has a substantial atmosphere. Its hazy clouds are too thick to see through, so what lies below is a mystery. Scientists think that Titan's air is like the air that existed on Earth billions of years ago, so this unusual moon might hold clues to the origin of life. Unfortunately, Titan's freezing surface is much too cold for life as we know it.

ARTIST'S IMPRESSION OF PROBE LANDING ON TITAN

MINI MOON

Apart from Titan, Saturn's moons are very small. Enceladus is only 500 km (310 miles) wide – it would easily fit inside France. Its freezing surface is covered with ice, but its interior is warm and probably contains liquid water. The grooves on its face were made by liquid water that welled up onto the surface and froze. Enceladus might have "ice volcanoes" that throw out snow when they erupt.

Saturn's rings are 100,000 km (62,000 miles) across but less than 1 km (0.6 miles) thick. If you made a pancake to the same proportions it would be 500 m (1,600 ft) wide.

THE VIEW OVER RHEA

Saturn's second-largest moon, Rhea, bears the scars of numerous meteorite and comet collisions. This artist's impression shows two of the largest craters in Rhea's icy surface. The larger crater, called Izanagi, is 230 km (143 miles) wide. Saturn dominates the skyline, and the distant Sun is small and feeble. Saturn's rings look paper-thin because they are seen edge-on.

URANUS

BLAND AND FEATURELESS is how many astronomers describe this planet. Even the *Voyager 2* probe saw nothing on its surface except some tiny clouds during a flyby in 1986. Astronomers classify Uranus as a "gas giant", although it mostly consists of a soup of hot water and other chemicals. A thick layer of gases covers the liquid interior. Uranus was discovered in 1781 by William Herschel, a German music teacher and amateur astronomer who had settled in England. Herschel wanted to call it "George's Star" in honour of King George III of England, but the name was later changed to Uranus, the Greek god of the heavens.

SIZE COMPARED TO EARTH

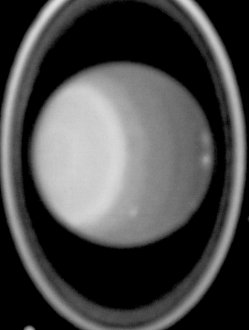

MAD MIRANDA

The strangest of Uranus's 17 moons is undoubtedly Miranda. This bizarre object has such a chaotic surface that scientists think it may have been smashed apart by an impact and then brought back together by gravity. Some parts of Miranda's core are now on the surface, while parts of the surface seem to have been buried. Miranda is named after a character in *The Tempest*, a play by William Shakespeare.

TOPSY TURVY

Uranus is tipped so far onto its side that it rolls along as it orbits the Sun instead of spinning upright. The moons and rings revolve around it like a giant Ferris wheel. The poles of Uranus take turns pointing at the Sun, resulting in some strange seasons. Each pole has 21 years of continual sunlight in summer and 21 years of continual darkness in winter. It is possible that Uranus was knocked into its tilted position by a massive collision billions of years ago.

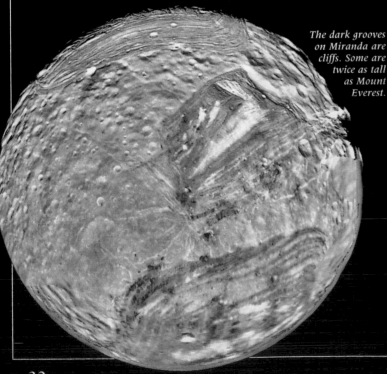

The dark grooves on Miranda are cliffs. Some are twice as tall as Mount Everest.

URANIAN RINGS

No one knew Uranus had rings until 1977, when astronomers caught sight of a star twinkling mysteriously as Uranus passed in front of it. The twinkling was caused by Uranus's rings blocking the star's light. *Voyager* took a closer look at the rings when it visited Uranus nine years later. It produced this false-colour image (right) of a slice through the ring system. Here, the rings appear white or pale, but they are really as black as charcoal.

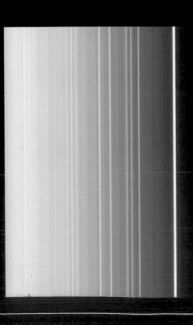

The Voyager *probe travelled for eight and a half years to reach Uranus. It then used Uranus's gravity to fling it on the final leg of its journey to Neptune.*

LAST LOOK

Voyager 2 caught this view of a crescent Uranus from 1 million km (620,000 miles) away as it left the planet. Methane gas in the atmosphere scatters blue light, giving the planet its colour. Uranus is 19 times further from the Sun than Earth so it gets extremely cold – the temperature on the cloud tops is –210°C.

NEPTUNE

IN THE OUTER REACHES of the Solar System is Neptune, a bitterly cold and dark world. Neptune is so far away that it would take over 200 years to get there at the speed of Concorde. Astronomers knew almost nothing about this planet until the *Voyager 2* space probe paid it a visit in 1989. When *Voyager* beamed its pictures to Earth, the astronomers were amazed. Despite its tranquil appearance, Neptune has the most violent weather in the Solar System, with storms the size of Earth and freezing winds ten times faster than a hurricane.

Voyager 2 has now left Neptune and is heading out of the Solar System.

SIZE COMPARED TO EARTH

GREAT GASBAG

Neptune is a giant globe of liquid and gas. Deep inside is an extremely hot, rocky core. This merges into a vast ocean of water and other chemicals that together make up most of the planet. The top of the ocean fades gradually into Neptune's "air" – a mixture of hydrogen and helium. The blue colour comes from a small amount of the gas methane. Neptune's upper atmosphere also contains a haze of complex organic chemicals, which shows up red in the picture above.

The dark area is a storm bigger than the Earth. After Voyager took this photograph, the storm disappeared.

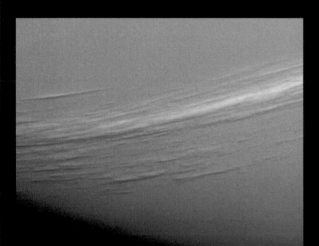

WINDS
OF CHANGE

Neptune's face is continually changing. Giant storms come and go, and glide around the planet carried by raging winds. The winds on Neptune can reach 2,000 kmh (1,240 mph) – the fastest in the Solar System. They sweep clouds of frozen methane into long white streaks. In the picture above, the clouds have cast shadows on the main deck of blue cloud 50 km (31 miles) below.

The temperature at Neptune's cloud tops is a freezing −220°C, but the planet's core is about 7,000°C – hotter than the surface of the Sun.

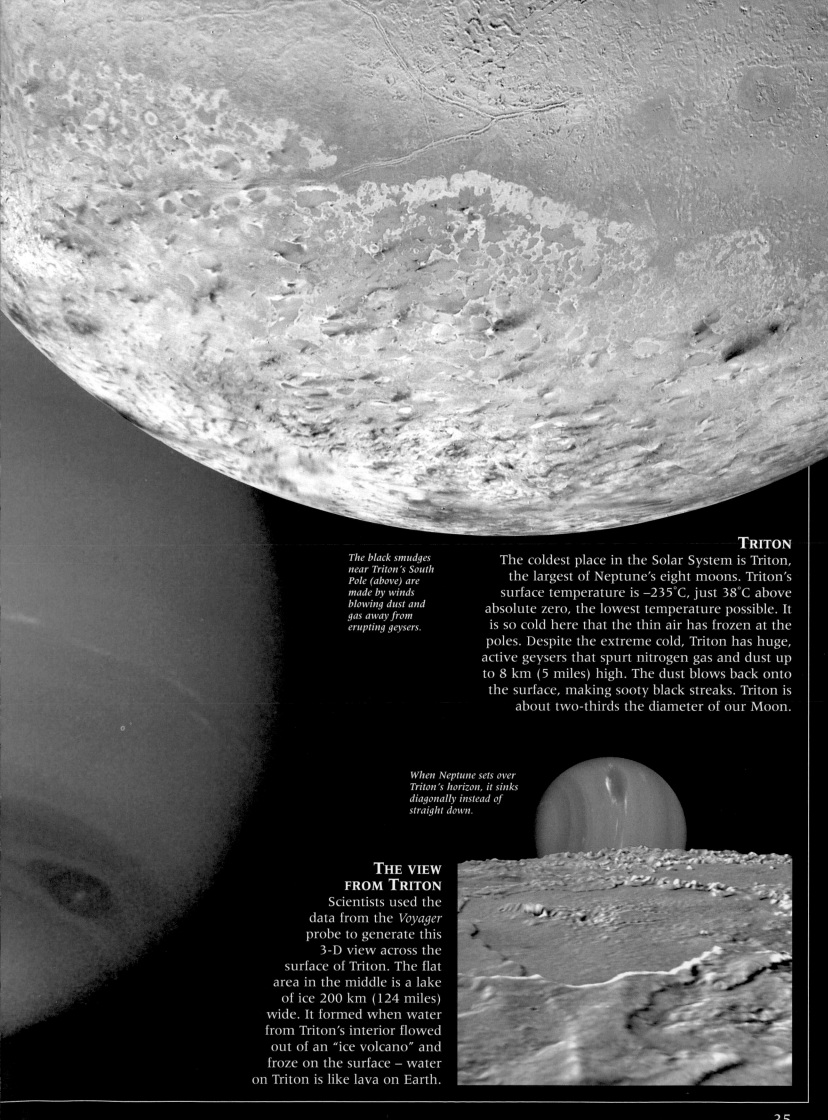

The black smudges near Triton's South Pole (above) are made by winds blowing dust and gas away from erupting geysers.

TRITON

The coldest place in the Solar System is Triton, the largest of Neptune's eight moons. Triton's surface temperature is –235°C, just 38°C above absolute zero, the lowest temperature possible. It is so cold here that the thin air has frozen at the poles. Despite the extreme cold, Triton has huge, active geysers that spurt nitrogen gas and dust up to 8 km (5 miles) high. The dust blows back onto the surface, making sooty black streaks. Triton is about two-thirds the diameter of our Moon.

When Neptune sets over Triton's horizon, it sinks diagonally instead of straight down.

THE VIEW FROM TRITON

Scientists used the data from the *Voyager* probe to generate this 3-D view across the surface of Triton. The flat area in the middle is a lake of ice 200 km (124 miles) wide. It formed when water from Triton's interior flowed out of an "ice volcano" and froze on the surface – water on Triton is like lava on Earth.

PLUTO & CHARON

MYSTERIOUS PLUTO IS THE SMALLEST and most distant planet. It is smaller than our Moon and 12,000 times further away, which makes it almost impossible to see. Looking at its surface from Earth is like trying to read the print on a postage stamp from 65 km (40 miles) away. Even so, astronomers have made some remarkable discoveries about Pluto and its moon, Charon. For instance, the pair make up a "double-planet system" – instead of Charon orbiting Pluto, they swing around each other. Pluto has the longest year of the planets, taking 248 Earth years to orbit the Sun. It has not even completed half an orbit since its discovery in 1930.

SIZE COMPARED
TO EARTH

SHADES OF GREY

Pluto is the only planet that has not been visited by a space probe, so astronomers have no clear pictures of its surface. The best photographs taken so far are these grey images of Pluto and Charon from the Hubble Space Telescope. They show mysterious bright and dark zones across both worlds. The bright areas at the poles may be icecaps.

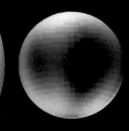

The strange dark and pale patterns on Pluto and Charon may be caused by frost or impact craters.

TWILIGHT WORLD

Astronauts on Pluto would have to carry torches all the time because it is always dark, even in the middle of the day. The Sun is so distant that Pluto gets very little warmth – the planet's surface is a chilly –230°C. In winter, Pluto's thin air freezes onto the ground. A day on Pluto, from sunrise to sunrise, lasts 153 hours.

PLUTO

Most of the Solar System's outer planets are made of gas and liquid, but Pluto is the exception. It is a solid ball of rock and ice, with only a very faint atmosphere. Pluto has an unusual, oval-shaped orbit around the Sun. It is the furthest planet from the Sun during its 200-year winter. However, during its short summer it moves nearer to the Sun than Neptune.

At more than half Pluto's width, Charon is the largest moon relative to its planet in the Solar System.

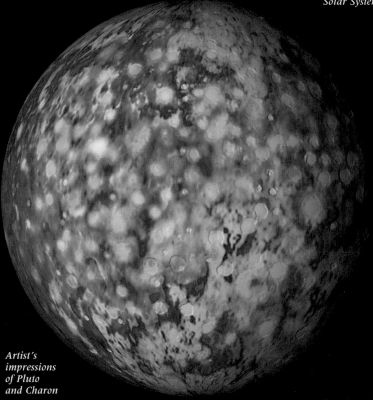

Artist's impressions of Pluto and Charon

CHARON

Unlike Pluto, Charon has a slight bluish tinge and may have water ice on its surface. This suggests that the two worlds formed in different places and were somehow brought together. Pluto and Charon keep their faces locked together. As a result, a visitor on Pluto would see Charon fixed in the same spot in the sky all the time. It would also look massive. Charon is so close to Pluto that it would appear six times wider than our own Moon does from Earth.

Pluto has such weak gravity that an astronaut who weighed 83 kg (182 lbs) on Earth would weigh only 3 kg (7 lb) on Pluto.

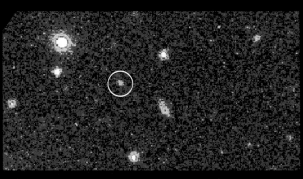

BEYOND PLUTO

Pluto is the furthest planet, but the Solar System is far from empty beyond it. Astronomers have recently begun to discover strange, icy objects lurking in the outer limits of the Solar System. The pictures on the left show one of them moving against distant stars. Called Kuiper objects, they are only a few hundred kilometres across, and are thought to turn into comets if they get too close to the Sun. Pluto and Charon may themselves be Kuiper objects – the largest of millions in the outer Solar System.

NASA is planning to send a small spacecraft to Pluto to arrive between 2010 and 2015. Called the Pluto–Kuiper Express, it will take about nine years to reach its target. It may carry a separate probe that will crash-land onto Pluto's surface.

COMETS & ASTEROIDS

Keen watchers of the night sky know that, from time to time, new objects appear against the familiar background of the stars. These newcomers are pieces of space debris, ranging in size from specks of dust to chunks of rock hundreds of kilometres wide. The most spectacular are the comets – balls of ice with dazzling tails that stretch for millions of kilometres. The largest pieces of space debris are massive rocks called asteroids. Most asteroids stay within an area called the asteroid belt, between the orbits of Jupiter and Mars. Sometimes, however, stray asteroids or comets end up on a collision course with a planet, with potentially catastrophic results.

Comets have two tails – a yellow or white tail made of dust and a blue tail made of gas.

COMETS

Comets have oval orbits that take them swooping close to the Sun and then flying far away. When a comet is near the Sun, its surface evaporates and releases gas and dust. The Sun's energy blows the gas and dust into two huge tails. The tails always point away from the Sun, no matter what direction the comet flies in. Large comets are spectacular when they pass near Earth. In 1997, millions of people saw the brilliant comet Hale-Bopp (above). It will pass Earth again in the 44th century.

HALLEY'S COMET

The most famous comet, Halley's Comet, flies past Earth every 76 years. Its visit in 1066 was recorded on a French work of art called the *Bayeux Tapestry*. In 1986, a small probe called *Giotto* flew into Halley's Comet and took this picture of its nucleus – a block of ice about 16 km (10 miles) long. Jets of gas and dust spurt out of the nucleus and stream around it, forming a vast white cloud called a coma.

DEEP IMPACT

This photograph shows a ball of fire larger than Earth exploding on the planet Jupiter. The explosion happened in 1992 when the comet Shoemaker-Levy 9 (SL9) smashed into Jupiter at more than 200,000 kmh (125,000 mph). SL9 had already been torn into a string of 20 fragments by Jupiter's massive gravity. As each fragment hit Jupiter, it threw up a new fireball.

METEORITES

Chunks of space debris that hit the ground are called meteorites. So far, no human is known to have been killed by a meteorite, although one killed a dog in Egypt in 1911. Barringer Crater in Arizona, USA, formed 50,000 years ago when a 10,000-tonne lump of iron hit the ground. Very rarely, something much bigger hits Earth. Many astronomers think that the extinction of the dinosaurs was caused by a comet 10 km (6 miles) wide that hit Central America 65 million years ago.

Barringer Crater in Arizona, USA, is 1.2 km (0.75 miles) wide.

SHOOTING STARS

Every day, Earth puts on about 550 tonnes in weight thanks to space dust. The dust particles, or meteors, smash into the atmosphere at terrific speeds. As they enter the air they burn up, often producing dramatic streaks of light across the sky called shooting stars. Shooting stars can be seen on any clear night, but the best time to see them is during one of the regular meteor showers that happen every year (*see also* p. 62).

Ida's surface is covered with craters.

ASTEROIDS

Asteroids are nothing more than giant boulders. This one, called Ida, is more than 52 km (32 miles) long. If Ida collided with Earth, it would destroy all forms of life except bacteria. Waves of fire would sweep over the land, destroying cities and forests. There would be tidal waves, earthquakes, and massive floods of lava. The debris blasted into the sky would blot out sunlight for decades. Fortunately, Ida is not heading this way. It is safely confined to the asteroid belt, 290 million km (180 million miles) away from us.

Ida (above) is so big that it even has its own tiny "moon", Dactyl (right).

STAR BIRTH

A STAR IS A HUGE GLOWING BALL of hydrogen gas, lit up by nuclear reactions in its core. Stars are continually forming and dying throughout the Universe. They are born inside giant gas clouds called nebulas. The gas in a nebula is about 25,000 million million times thinner than air on Earth. Even so, it has just enough weak gravity to make parts of the nebula shrink into blobs. As these get smaller, they heat up. Eventually, they get so hot and dense that nuclear reactions start inside them and they turn into stars. The heat and light from newborn stars can make nebulas glow in brilliant colours.

STAR-BIRTH NEBULAS

The lower part of the Trifid Nebula is red because new stars inside it are heating the gas and making it glow. The blue area is a vast, interstellar dust cloud lit up by nearby stars.

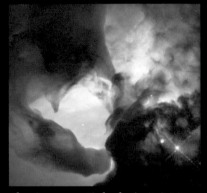

The Lagoon Nebula is 5,000 light years away so we see it as it was before the pyramids were built. This close-up shows a newborn star (red). Its radiation is blowing nearby clouds into strange shapes.

Not all nebulas are colourful. The Horsehead Nebula (right of centre) is a dense gas cloud that stands out against a red nebula behind it. It is probably on the brink of collapsing to form stars. The "head" is just the top part of an immense, dark cloud filling the lower right of this picture.

SEVEN SISTERS

Most new stars are born together in clusters.
Over billions of years, the stars may drift
apart and become solitary, like our Sun.
The Pleiades cluster is made up of massive,
white-hot stars that were born only 50
million years ago. They are so young that
they did not exist when dinosaurs roamed
the Earth. Pleiades are also called the
Seven Sisters because seven of the stars
are visible to the naked eye. Despite
the name, there are more than 500
stars in the cluster. The blue haze
around them is a cloud of dust.

STELLAR NURSERIES

These three colossal
pillars of hydrogen gas
are stellar nurseries.
Inside the tiny fingers
at the top of the tallest
pillar, hydrogen is
contracting and turning
into stars. Our Solar
System would easily fit
inside one of these
fingers. The gas pillars
lie 7,000 light years
away in the heart of the
Eagle Nebula. Heat from
nearby stars is slowly
boiling away the surface
of the pillars, producing
ghostly streamers of gas.
Eventually, the new
stars will emerge from
their cocoon of gas and
begin to shine.

*The tallest of these gas pillars
is about 10 million million km
(6 million million miles) in
height. New stars are forming
inside it, hidden by the dense gas.*

41

STAR DEATH

AFTER SHINING FOR MILLIONS OR BILLIONS of years, stars run out of fuel and die. The largest stars have short lives and violent deaths. As they age, they turn red and expand to a vast size – they become red supergiants. Finally, they self-destruct in an explosion brighter than a million suns. Smaller stars expand into stars called red giants as they age, but they live much longer and die quietly. Their remains are among the most beautiful objects in the Universe. Eventually, the debris from star death is recycled into new stars.

The Hourglass Nebula is a giant cloud made by a dying star as it disperses into space. The white spot in the "eye" is the star's collapsed core.

OLD-AGE SUPERSTAR

Antares is a red supergiant – a massive star nearing the end of its life. It has ballooned in size to about 1 billion km (600 million miles) wide – 700 times wider than our Sun. If the Sun was this big it would swallow all the planets out to Mars, including Earth. When Antares has used up all its fuel, it will blow itself apart in a supernova.

The red supergiant Antares is one of the brightest stars visible in the night sky.

A GALAXY AWAY, THIS STAR IS ABOUT TO DIE.

IT EXPLODES INTO A BRILLIANT SUPERNOVA.

The dying star Eta Carinae is spewing out gas and dust at over 2 million kmh (1.5 million mph).

SUPERNOVA

When a red supergiant runs out of energy, its core turns into iron. With no energy to hold its shape, the core collapses under its own gravity in seconds. The collapse sends out a blast wave that hurls the star's outer layers into space in an explosion called a supernova. Supernovas can outshine entire galaxies of stars. They are very rare and last for only a few months. In the biggest supernovas, the collapsing core crushes itself and turns into a black hole.

Eta Carinae is the only star known to emit natural laser light.

DEATH THROES

Massive Eta Carinae is perhaps the weirdest star known to science. Explosions are tearing it apart and flinging out gas and dust into a cloud of debris bigger than the Solar System. About 150 years ago, a particularly violent outburst made Eta Carinae the second brightest star in the sky. Today, it is impossible to see with the naked eye. Eta Carinae will probably destroy itself in a supernova explosion, but no one knows when. It could be tomorrow, it could be in a million years.

STAR GHOSTS

These ghostly shapes are glowing clouds of wreckage left by dead stars. Astronomers call them planetary nebulas. In the centre of a planetary nebula is all that remains of the original star's collapsed core – a tiny star called a white dwarf, which slowly cools and fades. White dwarfs are very dense – a single teaspoonful weighs half a tonne. The light they emit makes the gas and dust around them glow. Our Sun will produce a planetary nebula in about 5 billion years. Before then, it will turn into a red giant and swallow Earth.

This nebula formed when a dying red giant threw its outer layers into two massive wings.

Two expanding bubbles of gas have formed around this small white dwarf.

A thousand years ago, the Cat's Eye Nebula (right) was a red giant. The star's outer layers have now spread into shells of gas, lit by a central white dwarf.

43

THE MILKY WAY

O UR SUN BELONGS TO A GIGANTIC whirlpool of stars called the Milky Way. Huge families of stars are called galaxies, and like all galaxies, the Milky Way is unimaginably vast. It contains 200 billion stars. To count them all at one per second would take more than 6,000 years. Yet the Milky Way is mostly empty space. If you made a scale model of the Milky Way with a grain of sand for each star, the nearest star to the Sun would be 6 km (4 miles) away. The whole model would stretch a third of the way to the Moon. It would take millions of years to cross the Milky Way in a space shuttle.

HEART OF THE GALAXY

This picture from the Hubble Space Telescope shows stars packing the heart of the Milky Way galaxy. The stars have different colours depending on their age and size. Hot young stars are often blue, while large old stars are red. Astronomers think that a colossal black hole lies in the exact centre of our galaxy.

COSMIC WHIRLPOOL

Like everything else in space, the Milky Way is spinning around. This has spread the stars into a giant spiral. If we could see the Milky Way from outside, it would look like the galaxy shown above. Our Solar System lies halfway along one of the spiral arms and orbits the galactic centre once every 220 million years. The Milky Way is flat, so from Earth we see it side-on as a band of stars across the sky.

Seen from the side, the Milky Way is flat with a large central bulge.

STAR-BALLS

The oldest stars in the Milky Way are packed together in giant balls called globular clusters. The cluster above is Omega Centauri, a ball of several million stars. It contains stars as old as the Milky Way itself. Globular clusters orbit the centre of the galaxy in the same way as satellites go round the Earth.

On a very dark night, you can see the main disk of the Milky Way as a faint, glowing band across the sky. Every star visible to the naked eye is in the Milky Way.

SPACE DUST

Clouds of dust and gas thousands of times larger than our Solar System are interwoven with the stars of the Milky Way. Many are embedded with jewel-like stars that make the gas and dust glow, producing the most colourful sights in the galaxy. To the left is a view of multicoloured clouds near the star Antares (yellow, lower left), a close neighbour of our Sun.

GALACTIC NEIGHBOUR

The furthest thing you can see with the naked eye is not a star but another galaxy – Andromeda. Andromeda is one of the Milky Way's closest neighbouring galaxies. Its light takes two million years to reach us, so we see it as it was before humans walked on Earth. It is a spiral galaxy, but it looks flat because we see it from an angle. The two bright objects in this picture are smaller "dwarf galaxies".

GALAXIES

NEARLY ALL THE MATTER IN THE UNIVERSE is concentrated in galaxies. A galaxy is a gigantic mass of stars held together by gravity. The largest contain millions of millions of stars. The smallest have just a few million, but even small galaxies are so big that light takes thousands of years to cross them. Despite having so much matter, galaxies are mostly empty space, with vast distances between each star. Our Sun and all the stars we can see with the naked eye belong to just one galaxy – the Milky Way. Beyond this lie billions of galaxies stretching as far into space as astronomers can see.

ELLIPTICAL GALAXIES

Most galaxies are egg-shaped (elliptical). These galaxies are made up of masses of old, red stars that all formed around the same time. Elliptical galaxies have no gas for making new stars. The elliptical galaxy M87 (left) is the largest galaxy known. It contains 3 million million stars – 15 times as many as our Milky Way. Hidden in its centre is a massive black hole.

IRREGULAR GALAXIES

Galaxies with no recognizable shape are called irregular. They are usually small, with lots of young stars and bright gas clouds where new stars are forming. A typical example is the Large Magellanic Cloud (right). At 160,000 light years away, it is one of the closest galaxies and is visible to the naked eye as a pale smudge. It has only 10 billion stars – our Milky Way has 20 times more. The Large Magellanic Cloud is trapped by the Milky Way's gravity and orbits it every 6,000 million years. Eventually, the Milky Way's gravity will tear it apart and the two galaxies will merge.

Galaxies often contain colourful clouds of gas, such as the pink Tarantula Nebula (lower left).

The Sombrero Galaxy is 100,000 light years wide and contains 1,300 billion stars.

MEXICAN HAT GALAXY

The Sombrero Galaxy is really a spiral galaxy, but it looks like a Mexican hat because we see it from the side. The dark rim is a lane of dust that blocks out the light behind it. The Sombrero is an unusual type of galaxy. Its spiral arms are very hard to see because the central bulge is so large and bright.

SPIRAL GALAXIES

The most spectacular galaxies are spiral. These spin around like giant whirlpools, spreading their stars into graceful trailing arms. The oldest stars are located in a dense central hub. The spiral arms contain young stars, pink nebulas, and dark lanes of gas and dust. Spiral galaxies are disk-shaped, so they appear flat if we see them from the side. Our Milky Way is a spiral galaxy.

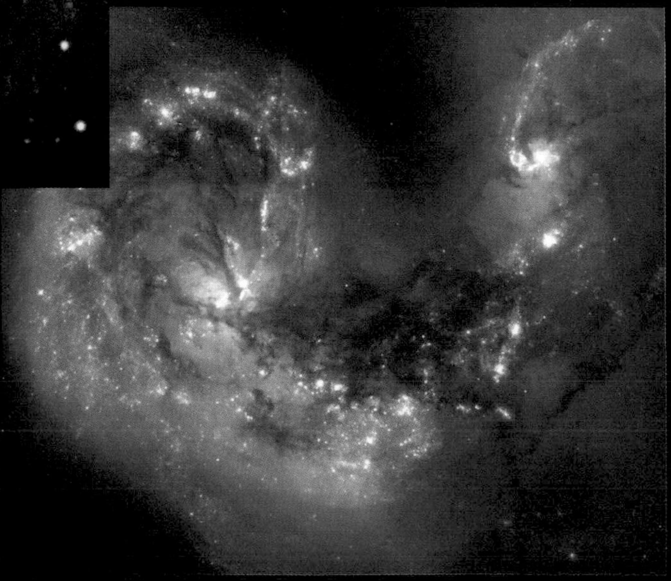

COSMIC COLLISION

Most galaxies are incredibly far apart, but some get close enough to collide. This picture shows two spiral galaxies crashing into each other. Their cores are orange. Individual stars do not collide, but dust clouds do, triggering a firestorm of star birth. The clusters of newborn stars in this picture look blue. The dark areas are dust clouds.

47

THE UNIVERSE

THE UNIVERSE IS EVERYTHING – planets, stars, galaxies, space, and even time. No one knows how big the Universe is – its size may even be infinite. Astronomers have discovered that all the galaxies in the Universe are flying away from each other, so the whole Universe must be getting bigger. But it is not expanding "into" anything because there is no space outside the Universe, and the Universe has no outer edge. The shape of the Universe is a mystery. It could be "curved", which means that if you travelled in a straight line for long enough, you would end up where you started. In some ways, the Universe is too strange to comprehend.

This picture shows a rippled pattern of radio waves coming from all directions in space. Scientists think it is the afterglow of the Big Bang.

BIG BANG
The Universe exploded out of nothingness about 13 billion years ago. Scientists call this violent beginning the Big Bang. No one knows what caused it. Time itself began with the Big Bang, so the question "what came before?" makes no sense. In a single second, the Universe grew from being smaller than an atom to being 20,000 light years wide. It is still expanding and may continue to do so forever.

OUR UNIVERSAL ADDRESS
Distances on Earth are dwarfed by the massive scale of the Universe. Imagine leaving your home on board a space shuttle to take a trip across the Universe. A few seconds after lift-off, your whole neighbourhood begins to shrink before your eyes...

HOME TOWN
Minutes later, your home town is visible in miniature. You can no longer make out separate houses, although you might just recognize the runways of an airport or the patchwork of fields in nearby countryside.

PLANET EARTH
Forty hours after launch, the whole Earth and Moon come into view. Swirling white clouds cover Earth, making the brownish patches of land difficult to see. There are no longer any obvious signs of human life.

THE SOLAR SYSTEM
After 27 years, you reach the edge of the Solar System. Along the way, you might pass Mars, Jupiter, Saturn, Uranus, and Neptune. The Sun – our local star – would always be visible.

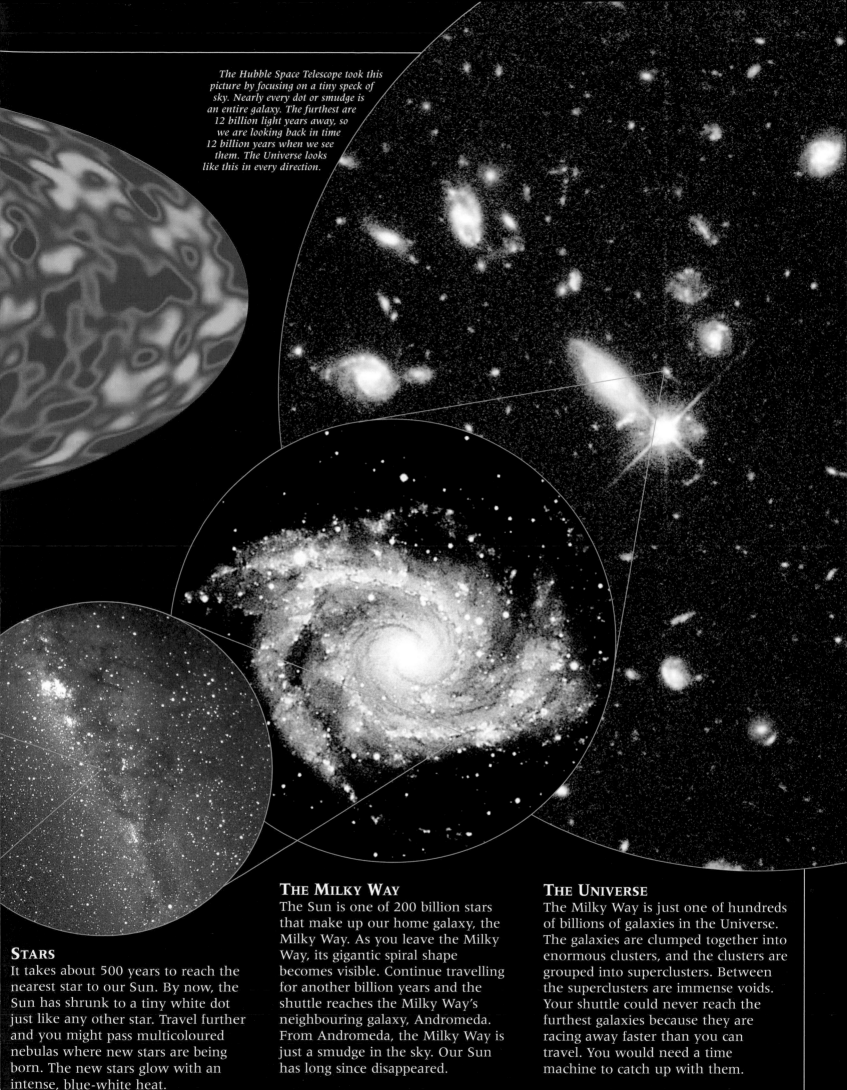

STARS

It takes about 500 years to reach the nearest star to our Sun. By now, the Sun has shrunk to a tiny white dot just like any other star. Travel further and you might pass multicoloured nebulas where new stars are being born. The new stars glow with an intense, blue-white heat.

THE MILKY WAY

The Sun is one of 200 billion stars that make up our home galaxy, the Milky Way. As you leave the Milky Way, its gigantic spiral shape becomes visible. Continue travelling for another billion years and the shuttle reaches the Milky Way's neighbouring galaxy, Andromeda. From Andromeda, the Milky Way is just a smudge in the sky. Our Sun has long since disappeared.

THE UNIVERSE

The Milky Way is just one of hundreds of billions of galaxies in the Universe. The galaxies are clumped together into enormous clusters, and the clusters are grouped into superclusters. Between the superclusters are immense voids. Your shuttle could never reach the furthest galaxies because they are racing away faster than you can travel. You would need a time machine to catch up with them.

ROCKETS

SPACECRAFT NEED THE MASSIVE POWER of rockets to launch them into space. Rockets work by burning fuel and then using the fast expulsion of exhaust gases to reach the speed and height they need. The most efficient way to achieve this is to use rockets divided into stages, generally three, one stacked on top of the other. The first stage lifts the vehicle into the high atmosphere. Once the fuel is expelled, the empty stage falls to Earth. The second, lighter stage takes over and reaches an even higher speed before falling away. The final stage puts the spacecraft into orbit or further into space.

ROCKET TO THE MOON

The flames of an *Apollo* launch light up the night sky. To escape Earth's gravitational pull and reach the Moon, its *Saturn V* rocket had to reach a speed of 40,000 kmh (24,855 mph) – known as Earth's escape velocity. The world's most powerful rocket, it gulped down 13 tonnes of fuel a second and was as tall as a 32-storey building. It was used for the last time in 1973.

VOSTOK I

On 12 April 1961, the Soviet cosmonaut Yuri Gagarin made history when he became the first person in space. The *Vostok 1* rocket, which carried his capsule into orbit, was a modified version of a military missile. Its design was kept secret for years. After successfully completing one orbit of the Earth (108 minutes), Gagarin bailed out at 6,100 m (20,000 ft) and parachuted into a Russian field.

SOYUZ

Today, Russian cosmonauts fly into space on top of a *Soyuz* rocket – a modern version of Gagarin's *Vostok 1*. *Soyuz* has four cone-shaped boosters at its base that are used to add to the thrust. The crew's capsule lies beneath the white section, shown in the photo, and a small emergency rocket on top can pull the astronauts free if the launch goes wrong. *Soyuz* is mainly used to ferry supplies and crew from Earth to space stations, such as *Mir*.

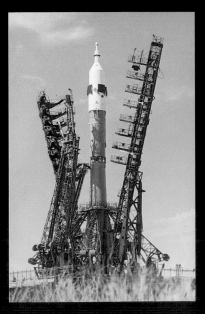

SPLASHDOWN

Getting back to Earth is a risky business.
When the *Apollo* command module returned
to Earth it travelled faster than humans had
ever gone before. As it entered the Earth's
atmosphere it started to burn up like a
shooting star. The module (the only part
of *Apollo* to return) was so heavy that it
needed three parachutes to slow it down
before it hit the water. Soon after impact,
inflatable air bags turned the module
upright and prevented it from
sinking. The astronauts were taken
to safety, and a helicopter took the
charred module to a waiting ship.

ARIANE

Europe's *Ariane* rockets are
used to put communications
satellites or space probes into
orbit. This cargo is called a
payload. The latest of these
rockets is *Ariane 5*. Its main
body is flanked by two
massive rocket boosters, each
with a parachute for a slow
descent into the ocean. All
but the payload will fall to
Earth or burn up in the
atmosphere. The
rockets lift off from
French Guiana in
South America.

ROCKETS OF THE FUTURE

The next generation of launchers will be
space planes, not rockets. They will use their
own engines to escape Earth's gravitational
pull. This means they will be completely
reusable and much cheaper to run. One such
design is a space plane called *VentureStar*. This
will lift off from an
upright position
but land belly
down on a runway.
VentureStar may carry
cargo and crew to
the *International Space
Station* after its completion
in the 21st century. A
half-size prototype,
the X-33, should start
sub-orbital test flights
in 2000.

51

THE SPACE SHUTTLE

THE SHUTTLE IS A REMARKABLE piece of engineering. It is the only spacecraft that can fly into space and survive the return to Earth to fly again. The reusable part of the Shuttle is called the orbiter. When the orbiter is joined to its orange fuel tank and two rocket boosters, the whole shuttle weighs about 2,000 tonnes. It takes the largest land vehicle in the world to move them to their launch pad in Cape Canaveral in Florida. An average shuttle mission lasts about 10 days. During that time, the orbiter circles Earth about 160 times.

Technicians and crew reach the Shuttle an aerial walkway leading from the launch tower.

LIFT-OFF!

A trail of flame bursts from the rocket boosters as the Shuttle clears the tower. Like giant fireworks, once they ignite they cannot be shut off until they run out of fuel. The three main engines use liquid fuel from the giant orange tank. During their eight-minute run, they drink 700 tonnes of fuel and produce enough power to light the state of New York.

Main fuel tank

Rocket booster

Orbiter

INTO ORBIT

It takes only eight and a half minutes for the Shuttle to reach space. As it accelerates towards orbit, the crew feels three times heavier than normal. Once the boosters and fuel tank have separated, the orbiter's own engines put it into orbit. When the orbiter reaches an altitude of about 320 km (200 miles), it is travelling at 28,000 kph (17,500 mph).

BURNING UP

The largest section of the Shuttle is the main fuel tank, which is as tall as a 15-storey building. Before the Shuttle reaches orbit the empty fuel tank is released and burns up during re-entry into the atmosphere. The solid boosters also fall away shortly after lift-off. They parachute into the ocean and are recovered to be used on another mission.

THE CHALLENGER DISASTER

Seven astronauts died at the start of the 25th Shuttle mission on 28 January 1986. *Challenger* exploded 73 seconds after launch. Hot gas leaking from one of the solid rocket boosters burned through the main fuel tank and set light to the fuel. The Shuttle's remains fell into the sea.

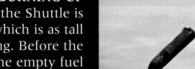

NASA
Atlantis

USA

MISSION CONTROL

As soon as the Shuttle lifts off from Cape Canaveral, the Mission Control Center in Houston, Texas, takes charge. During missions, the control centre operates 24 hours a day. The flight controllers work in three shifts. About 20 flight controllers in each team carefully ▮▮▮ information from the Shuttle on their computer screens. A large map of the world shows the Shuttle's location, while a nearby video screen shows live TV broadcasts.

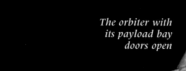

The orbiter with its payload bay doors open

SPACE GARAGE

Satellites, experiments, and laboratories can be stored in the Shuttle's payload bay. On some missions, there is an extra cabin in the payload bay. A tunnel links the cabin to the living quarters in the front of the orb▮▮. In this picture, two astronauts give the Hubble Space Telescope its regular service. They are standing on the end of a moving arm attached to the payload bay.

As the astronauts work, the Florida coastline can be seen far below.

ANATOMY OF A SHUTTLE

Most of the orbiter's bulk is taken up by the payload bay. Five cars could fit end-to-end in this area. The crew of up to eight squeeze into the working and living quarters in the front section of the orbiter. The front section is split into two rooms. On top is the flight deck, where the commander and pilot control the orbiter. Below the flight deck is the mid-deck, where crew members eat, sleep, wash, and exercise.

After hitting the runway at around 320 kmh (200 mph), a parachute is used to slow down the orbiter.

GLIDING TO EARTH

The orbiter becomes the world's largest glider when it returns to Earth. Friction with the air slows it down as it re-enters Earth's atmosphere, and special tiles protect it from searing temperatures outside. Five computers control the orbiter during re-entry, then the commander takes over to land the craft. If it misses the runway, it is impossible to turn it around and try again. If the orbiter lands in California, it has to be carried back to Florida on the back of a specially adapted jumbo jet.

LIVING IN SPACE

WEIGHTLESSNESS IS THE STRANGEST thing about living in space. Actions that were impossible on Earth suddenly become easy. Astronauts can turn effortless somersaults in mid-air and can even lift giant satellites over their heads. But zero gravity can also make life difficult. When a crew member needs to shave, he must vacuum up the cuttings before they start floating around. To stay in one place, astronauts must secure themselves in footholds. They also have to change the way they eat, drink, sleep, and use the toilet. At first, many crew members suffer from space sickness, but after several days, their bodies adapt.

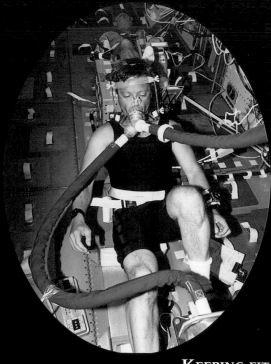

SLEEPING ON THE WALL
There is no up or down in space, so astronauts can sleep on a wall or on the ceiling. Some people use bunks, while others attach sleeping bags to a wall (right) so as not to float away. There is no normal day and night in space so sleeping astronauts have to wear eyeshades.

KEEPING FIT
With no gravity to push against, muscles and bones become weak. Astronauts fight this by exercising every day. They may walk on a treadmill, use a rowing machine, or pedal on an exercise bicycle (above). Most people "grow" about 5 cm (2 in) during their time in space because in zero gravity their spines stretch.

DRINKS

FRUIT AND NUTS

CEREALS

SWEET AND SOUR BEEF

READY MEALS
Space crews eat three meals a day. They can choose from more than 70 hot and cold meals and several types of drink. Each person's food is marked with a coloured dot. Many foods, such as sausages, are similar to those found in supermarkets. Others need to be mixed with hot water. Snacks and sweets are also available. All the menus are planned before the mission.

When drinks escape from their containers they float around as giant bubbles.

Drinking straw

Astronauts wear normal clothes inside the Space Shuttle.

FLOATING FREE
Meal times can be crowded on the Space Shuttle's mid-deck (above). Staying in one place is a problem but astronauts are happy to eat in any position. Packets of food are stuck to lockers to prevent them floating away. Liquids would not stay in a cup so drinks come in plastic packets with straws.

WORKING IN SPACE
When satellites break down, astronauts may be able to rescue them. Here, an astronaut is capturing a communications satellite that was stuck in the wrong orbit. On Earth it would have weighed several tonnes. The satellite was later returned to Earth to be repaired.

Inside the helmet is a microphone and headphones for talking to other astronauts.

The backpack is like a flying armchair that allows the astronaut to move through space.

TV camera

The spacesuit has a special layer to keep pressure on the astronaut's body.

SPACE WALKING
A trip outside a spacecraft is called a space walk. During a space walk, the astronaut must wear a spacesuit to protect him from the airless vacuum of space. Without the suit, his blood would boil and he would die in seconds. Spacesuits are like mini-spacecraft with their own power, oxygen, and cooling systems. Shuttle astronauts can also wear special backpacks that allow them to fly around freely.

Controls to steer astronaut in any direction

SUCTION TOILET
Toilets in space work like vacuum cleaners. Flowing air sucks waste into a container. On the *Mir* space station, the waste fluids are recycled into drinking water. On early space missions, the crew simply used waste bags or wore nappy-like underwear.

SPACE STATIONS

ASTRONAUTS CAN SPEND months or years in space by living on space stations. The climate in a space station is carefully controlled, and stale air and dirty water are constantly recycled. Huge solar panels provide an endless supply of power. The record for the longest period spent in space is held by a Russian doctor called Valeri Poliakov, who spent 438 days on space station *Mir*. Thanks to Poliakov and other space station veterans, scientists now know more about the ways zero gravity affects the human body. This knowledge will be vital if astronauts ever set off to explore other planets.

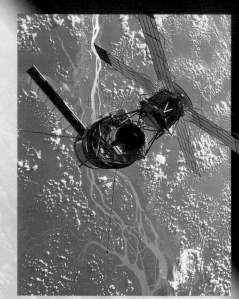

SKYLAB

The US space station *Skylab*, seen here above the Amazon rainforest, was made of leftovers from the *Apollo* Moon mission – the living area was the top part of a *Saturn* rocket. Nine astronauts lived on *Skylab* between 1973 and 1974. They carried out hundreds of experiments. The space station was then abandoned and burned up as it fell back into the atmosphere in 1979.

MIR

The Russian space station *Mir* has been in orbit since 1986. It was built in space by joining together different sections, called modules, one at a time. *Mir* has had a troubled history. There have been two fires on board, and in 1997 one of the modules was wrecked by a crash with a supply craft. The station is now nearing the end of its life. Eventually, its crew will abandon *Mir* and allow it to burn up as it falls back to Earth.

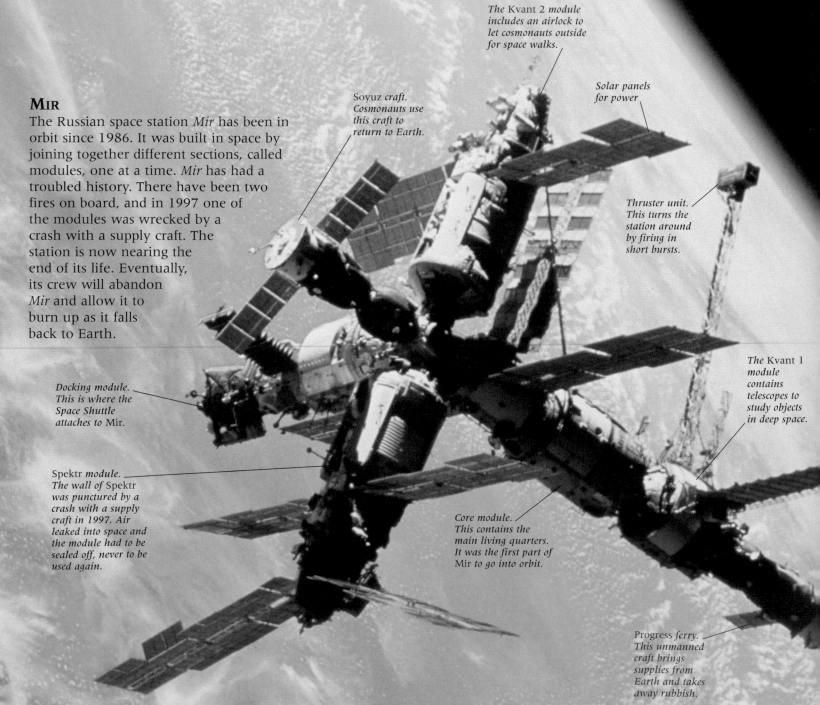

The Kvant 2 module includes an airlock to let cosmonauts outside for space walks.

Soyuz *craft. Cosmonauts use this craft to return to Earth.*

Solar panels for power

Thruster unit. This turns the station around by firing in short bursts.

The Kvant 1 module contains telescopes to study objects in deep space.

Docking module. This is where the Space Shuttle attaches to Mir.

Spektr *module. The wall of Spektr was punctured by a crash with a supply craft in 1997. Air leaked into space and the module had to be sealed off, never to be used again.*

Core module. This contains the main living quarters. It was the first part of Mir to go into orbit.

Progress *ferry. This unmanned craft brings supplies from Earth and takes away rubbish.*

EAST MEETS WEST

As a sign of greater friendship between Russia and the USA, shuttles began to visit *Mir* in 1995. To reduce the risk of a disastrous crash, the shuttles had to slow down to 0.1 kmh (0.06 mph) as they docked. The shuttle astronauts then floated through a tunnel and into *Mir* to meet the Russian cosmonauts. Each shuttle visit also brought fresh supplies and experiments for the *Mir* crew.

Seven astronauts and three cosmonauts (Russian astronauts) get head to head during a party on one of the shuttle visits to Mir.

Space shuttle Atlantis *docking with Mir in 1995*

INSIDE MIR

Conditions are cramped and messy in *Mir*. It is difficult for the cosmonauts to stay clean, so *Mir* is also very smelly. Almost every surface in the core module (right) is covered with equipment, and tangled wires float in the air. Although there is no up or down in space, some rooms have carpeted "floors" and "ceiling" lights to make the crew feel at home. Two crew members have private cabins with portholes and sleeping bags. The rest sleep in the main modules.

INTERNATIONAL SPACE STATION

Countries around the world are co-operating to build the *International Space Station (ISS)* – the world's largest space station. The *ISS* will take at least six years to build at an estimated cost of 40 billion dollars. At twice the size of a football pitch, it will be the brightest "star" in the sky. The first module was launched on 20 November 1998. Forty-four more flights will deliver the remaining pieces.

ESCAPE POD

Life in space is dangerous, so the *International Space Station* will have an escape craft to carry the crew of seven back to Earth in emergencies. The craft will be a type of shuttle that flies without a pilot. After entering the atmosphere, it will parachute to the ground and skid to a stop.

The Mir space station orbits the Earth once every 90 minutes at a height of 320 km (200 miles). It passes over most of the densely populated parts of the world, including Europe and North America.

SATELLITES

A SATELLITE IS ANYTHING that orbits a planet. Our Moon, for example, is a natural satellite. Earth also has about 500 artificial satellites – machines that carry out many automatic tasks, from watching the weather to transmitting TV pictures. Satellites are launched into space by rockets or the Space Shuttle and are released at the correct height to stay in orbit. They are usually powered by large solar panels that catch the intense sunlight in space. Most satellites orbit at the same speed that Earth rotates, so they always stay over exactly the same spot.

SPIES IN THE SKY
The most advanced satellites are used for spying. Some have telescopes so powerful that they can read newspaper headlines from space. Here, a US spy satellite is being released from a space shuttle. This satellite gives early warnings of missile attacks by detecting heat from missile exhausts.

Sputnik 1 consisted of a metal ball 58 cm (23 in) wide with four aerials. Inside the ball was a radio transmitter.

SPUTNIK
In October 1957, the Soviet Union astonished the world by launching the first ever satellite into space. *Sputnik 1* hurtled around Earth every 96 minutes, beaming down a "bleep bleep" signal. All over the world, people looked for it in the night sky. After 92 days, it fell into the atmosphere and burned up. Later, the Soviets launched *Sputnik 2*, which carried a dog called Laika. She died during the flight.

WEATHER WATCH
Weather satellites keep a constant eye on changing cloud patterns. The pictures they beam back help scientists forecast the weather, days or weeks in advance. In hurricane zones, such as the Caribbean, these forecasts can save lives. The satellite picture above shows Hurricane Fran heading towards the coast of North America in 1996. Florida lies to its left. Fran hit North Carolina and devastated more than 30,000 homes.

"LIVE BY SATELLITE!"
When you telephone someone on the other side of the world, a network of communications satellites (comsats) beams your voice through space at the speed of light. Comsats also broadcast live TV pictures. Just one comsat can transmit 40,000 telephone calls and several TV channels at the same time.

Communications satellites are about three times bigger than a car. They stay over the same place on Earth all the time.

MAPPING EARTH

Some satellites specialize in taking photographs of land. Their pictures have all sorts of uses. They can be used to make maps, to search for precious minerals, to monitor crops, or to study the effects of natural disasters like floods and forest fires. This satellite picture (right) shows San Francisco Bay in North America. The city of San Francisco lies at the northern end of the bay. Golden Gate Bridge is just north of the city. Alcatraz Island, once a famous prison, is to the city's northeast.

Golden Gate Bridge

Alcatraz Island

San Francisco

Mountains

Snow

Main road

Airport

Solar panels

AROUND THE WORLD IN 26 DAYS

The *SPOT* satellite (above) flies from pole to pole as it orbits Earth, photographing everything directly below. Because the Earth rotates, *SPOT* passes over different places with each successive orbit. So every 26 days, *SPOT* photographs our planet's entire surface. *SPOT*'s telescope is not as good as a spy satellite's – it can only just see the pyramids in Egypt.

Satellites send their pictures back to Earth as a stream of electronic data. Powerful computers then process the data to highlight particular features. For instance, the picture above has been processed to look natural. The version on the left has been processed to highlight vegetation (red areas).

IS ANYONE OUT THERE?

SO FAR, THE ONLY PLACE in the universe known to support life is Earth. However, our galaxy probably contains billions of planets that we cannot see, and there are billions of galaxies in the Universe. Many astronomers think it is very unlikely that Earth could be the only planet that has life. There is even a faint possibility that life might exist within our Solar System, under the icy surface of Jupiter's moon Europa, for example. Whether we will ever discover intelligent life forms is another matter. If alien civilizations do exist, they are probably too far away to visit.

EAVESDROPPING ON ALIENS

If alien civilizations exist, perhaps they use radio waves to communicate or to broadcast TV and radio programmes. If so, we might be able to "eavesdrop". Astronomers involved in SETI (Search for ExtraTerrestrial Intelligence) use powerful radio telescopes to scan the sky for alien signals. Apart from some false alarms, SETI astronomers have so far found nothing. Meanwhile, our own radio and TV signals are spreading into space and may one day give us away.

MARTIAN MICROBES?

In 1996, NASA scientists shocked the world with a remarkable discovery. They had found what looked like the remains of tiny life forms in a meteorite from Mars. The worm-like structures were 100 times thinner than a hair. However, after closer inspection, many scientists decided these "fossils" were just patterns in the rock.

ARECIBO RADIO TELESCOPE

If aliens ever stumble across the Pioneer or Voyager space probes, they will discover these messages from Earth. But will they understand them?

GOLD DISKS

The *Voyager* space probes carry old-fashioned records with recordings of Earth sounds, from Beethoven's music to greetings in 56 languages. More than 100 photographs are encoded on the records. Each probe also has a stylus for playing the records.

SEARCHING FOR ALIEN WORLDS

There may be billions of planets in our galaxy that support life, but modern telescopes are not powerful enough to see any of them. To search for new worlds beyond the Solar System, astronomers hope to build new telescopes. The telescopes would work from space, orbiting Earth as Hubble does.

Design for a future space telescope

MESSAGE IN A BOTTLE

As you read this, four space probes are heading out of our Solar System and into deep space. Each carries a message for any aliens that happen to find them. The two *Pioneer* probes carry engraved metal plates showing a man and woman and Earth's position in the Solar System. The two *Voyager* probes carry records that are gold-plated to protect them from space dust.

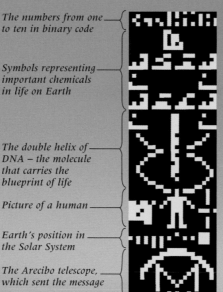

The numbers from one to ten in binary code

Symbols representing important chemicals in life on Earth

The double helix of DNA – the molecule that carries the blueprint of life

Picture of a human

Earth's position in the Solar System

The Arecibo telescope, which sent the message

THE ARECIBO MESSAGE

In 1974, scientists used the Arecibo radio telescope (below) to send a radio message into space in the hope that intelligent aliens will detect it. The message translates into some simple pictures (left) about life on Earth. It was beamed towards a cluster of half a million stars called M13. Unfortunately, M13 is so far away that we will have to wait at least 48,000 years for a reply, if one ever comes.

The M13 star cluster

ARECIBO

Arecibo has the largest telescope dish in the world. It is 305 m (1,000 ft) wide, and lies in a natural bowl in the hills of Puerto Rico. The curved dish reflects radio waves onto a detector/transmitter held overhead. The detector can be moved to aim the telescope in different directions.

SPACE DATA

SOLAR SYSTEM

	Sun	Mercury	Venus	Earth	Mars	Jupiter	Saturn	Uranus	Neptune	Pluto
Diameter (km)	1,392,000	4,878	12,103	12,756	6,786	142,984	120,536	51,118	49,528	2,274
Mass (Earth=1)	332,946	0.05	0.82	1	0.11	318	95.18	14.5	17.14	0.002
Surface gravity (Earth=1)	27.9	0.38	0.91	1	0.38	2.36	0.92	0.89	1.12	0.07
Average surface temperature (°C)	5,500	179	482	15	−63	−121	−180	−193	−200	−230
Daylength (rotation period in hours)	610–864	1,408	5,832	24	25	10	10	18	19	153
Length of a year (Earth days)	–	88	225	365	687	4,333	10,760	30,685	60,190	90,800
Distance from Sun (million km)	–	58	108	150	228	778	1,427	2,871	4,498	5,906
Speed of movement around Sun (kmh)	–	172,332	126,072	107,244	86,868	47,016	34,812	24,516	19,548	17,064
Number of rings	0	0	0	0	0	3	7	10	6	0
Number of known moons	–	0	0	1	2	16	18	17	8	1

TOTAL SOLAR ECLIPSES UNTIL 2010*

Date	Where visible
11 August 1999	India, Middle East, Europe, Atlantic
21 June 2001	Indian Ocean, S. Africa, S. Atlantic
4 December 2002	Australia, Indian Ocean, S. Africa
23 November 2003	Indian Ocean, Antarctica
8 April 2005	Central America
29 March 2006	Central Asia, Africa
1 August 2008	China, Russia, Greenland
22 July 2009	Pacific, China, India
11 July 2010	S. America, Pacific

* See also map on page 11

TOTAL LUNAR ECLIPSES UNTIL 2010

Date	Where visible
21 January 2000	N., S., and Central America, W. Europe, W. Africa
16 July 2000	Pacific, Australia, S.E. Asia
9 January 2001	Africa, Europe, Asia
16 May 2003	S. America, Antarctica
9 November 2003	S. America, Europe, W. Africa
4 May 2004	Africa, Middle East, India
28 October 2004	N., S., and Central America, W. Africa, W. Europe
3 March 2007	Africa, Europe, Middle East
28 August 2007	Alaska, North America, Pacific, E. Australia,
21 February 2008	N. America, S. America, W. Europe, W. Africa
21 December 2010	N. America, Pacific

ANNUAL METEOR SHOWERS

Name	Peak dates
Quadrantids	3–4 January
Lyrids	20–22 April
Eta Aquarids	4–6 May
Southern Delta Aquarids	28–29 July
Northern Delta Aquarids	6 August
Southern Iota Aquarids	6–7 August
Perseids	12 August
Northern Iota Aquarids	25–26 August
Orionids	21–22 October
Taurids	3–5 November
Leonids*	17–18 November
Geminids	13–14 December
Ursids	23 December

* May produce a spectacular meteor storm in 1999

COMPARATIVE SIZES

These photographs are shown to scale and illustrate the relative sizes of the Sun and all nine planets. The orange background on this page represents the Sun. The smallest planet, Pluto, is only the size of a pinhead.

MERCURY VENUS EARTH MARS

DISTANCE FROM SUN

The line below shows the distance of each planet from the Sun. Mercury is the closest. Pluto is more than 100 times further away.

JUPITER

SATURN

SUN MERCURY VENUS EARTH MARS   JUPITER   SATURN   URANUS

DISTANCE IN MILLIONS OF KILOMETRES

| 0 | 500 | 1,000 | 1,500 | 2,000 | 2,500 |

NEAREST STARS

Name	Constellation	Distance (light years)*
Sun	A–	0.000015
Proxima Centauri	Centaurus (The Centaur)	4.2
Alpha Centauri A	Centaurus (The Centaur)	4.3
Alpha Centauri B	Centaurus (The Centaur)	4.3
Barnard's Star	Ophiuchus (The Serpent Bearer)	5.9
Wolf 359	Leo (The Lion)	7.6
Lalande 21185	Ursa Major (The Great Bear)	8.1
Sirius A	Canis Major (The Great Dog)	8.6
Sirius B	Canis Major (The Great Dog)	8.6
UV Ceti A	Cetus (The Whale)	8.9

BRIGHTEST STARS FROM EARTH

Name	Constellation	Distance (light years)*
Sun	–	0.000015
Sirius A	Canis Major (The Great Dog)	8.6
Canopus	Carina (The Keel)	313
Alpha Centauri A	Centaurus (The Centaur)	4.3
Arcturus	Boötes (The Herdsman)	36
Vega	Lyra (The Lyre)	25
Capella	Auriga (The Charioteer)	42
Rigel	Orion (The Huntsman)	773
Procyon	Canis Minor (The Little Dog)	11
Achernar	Eridanus (River Eridanus)	144
Betelgeuse	Orion (The Huntsman)	427

* One light year = 9.46 million million km

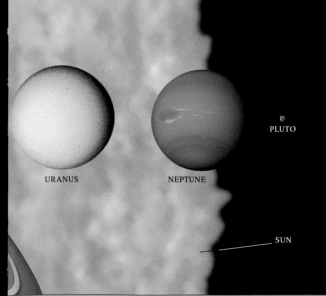

URANUS   NEPTUNE   PLUTO   SUN

3,500   4,000   4,500   5,000   5,500   5,900

NEPTUNE   PLUTO

LANDMARKS IN SPACE EXPLORATION

4 OCTOBER 1957 – Launch of *Sputnik 1*, the world's first artificial satellite.

3 NOVEMBER 1957 – Soviet dog Laika becomes first living creature in space.

31 JANUARY 1958 – Launch of *Explorer 1*, first US satellite.

10 OCTOBER 1959 – Soviet *Luna 3* spacecraft returns first pictures of Moon's far side.

LAIKA

12 APRIL 1961 – Soviet cosmonaut Yuri Gagarin becomes the first person in space.

5 MAY 1961 – Alan Shepard becomes first US astronaut to fly in space.

20 FEBRUARY 1962 – John Glenn becomes first US astronaut to orbit the Earth.

10 JULY 1962 – Launch of *Telstar 1*, world's first commercial communications satellite.

JOHN GLENN

16 JUNE 1963 – Soviet cosmonaut Valentina Tereshkova becomes first woman in space.

18 MARCH 1965 – First space walk (by Soviet cosmonaut Alexei Leonov).

15 JULY 1965 – US spacecraft *Mariner 4* completes first successful Mars flyby.

3 FEBRUARY 1966 – Soviet *Luna 9* becomes first craft to land successfully on the Moon.

27 JANUARY 1967 – Three US astronauts die in a fire on the launch pad of *Apollo 1*.

24 APRIL 1967 – First space death. Soviet Vladimir Komarov dies when his capsule's parachutes tangle on returning to Earth.

18 OCTOBER 1967 – Soviet *Venera 4* becomes first craft to land on Venus.

24 DECEMBER 1968 – *Apollo 8* becomes first craft to orbit the Moon carrying people.

20 JULY 1969 – US astronaut Neil Armstrong becomes first person to walk on the Moon.

17 NOVEMBER 1970 – Russian *Lunokhod 1* becomes first rover to drive on Moon.

LUNOKHOD 1

19 APRIL 1971 – Launch of Soviet *Salyut 1*, the world's first space station.

29 JUNE 1971 – Three Russians die as *Soyuz 11* capsule re-enters atmosphere.

19 DECEMBER 1972 – Splashdown of *Apollo 17*, the last mission to land astronauts on the Moon.

5 DECEMBER 1973 – US spacecraft *Pioneer 10* makes first flyby of Jupiter.

29 MARCH 1974 – US spacecraft *Mariner 10* makes first flyby of Mercury.

17 JULY 1975 – First time a US and Soviet spacecraft dock together (*Apollo–Soyuz*). Astronauts shake hands.

20 JULY 1976 – US spacecraft *Viking 1* becomes first craft to land successfully on Mars.

1 SEPTEMBER 1979 – US spacecraft *Pioneer 11* makes first flyby of Saturn.

12 APRIL 1981 – Launch of the first US Space Shuttle, *Columbia*.

24 JANUARY 1986 – US spacecraft *Voyager 2* makes first flyby of Uranus.

28 JANUARY 1986 – US Space Shuttle *Challenger* explodes during launch, killing the crew of seven.

20 FEBRUARY 1986 – Launch of Soviet *Mir* space station.

14 MARCH 1986 – European *Giotto* probe makes first close flyby of a comet (Halley's).

SPACE SHUTTLE COLUMBIA

25 AUGUST 1989 – *Voyager 2* makes first flyby of Neptune.

24 APRIL 1990 – Launch of the Hubble Space Telescope from Space Shuttle *Discovery*.

22 MARCH 1995 – Russian cosmonaut Valeri Poliakov returns after a record 438 days in space.

29 OCTOBER 1998 – US astronaut John Glenn becomes oldest ever space traveller at the age of 77.

20 NOVEMBER 1998 – Launch of first part of the *International Space Station*.

SPACE WEBSITES

Subject	Website address
Deep-space missions	http://www.jpl.nasa.gov
European Southern Observatory	http://www.eso.org/outreach
European Space Agency	http://sci.esa.int
Florida Today space news	http://www.flatoday.com/space
Hubble Space Telescope	http://www.stsci.edu
International Space Station	http://station.nasa.gov/station
Japanese space programme	http://www.nasda.go.jp
Mars missions	http://mars.jpl.nasa.gov
NASA photos of planets	http://photojournal.jpl.nasa.gov
Astronomy Now magazine	http://www.astronomynow.com
Space Shuttle	http://www.shuttle.nasa.gov
Views of the Solar System	http://bang.lanl.gov/solarsys/eng/homepage.htm

INDEX

ACKNOWLEDGMENTS

Special thanks to Cheryl Gundy at the Space Telescope Science Institute, Baltimore, and Debbie Dodds at Johnson Space Flight Center. Thanks also to Lynn Bresler for the index; Fran Jones, Amanda Rayner, and Sue Leonard for editorial assistance; Lester Cheeseman and Eun-A Goh for design assistance; Andrew O'Brien for DTP assistance; Chris Branfield for jacket design; and Robin Hunter for computer wizardry.

Dorling Kindersley would like to thank the following for their kind permission to reproduce their photographs.

(a=above; b=below; c=centre; l=left; r=right; t=top)

Anglo Australian Observatory: 4r, 5bl, 42l, 42bl, 46c, 46cl, 47r; David Malin 2tr, 40cl, 44cr, 44l, 49c; **Caltech:** Palomar Observatory 4bl; **Corbis UK Ltd:** Lowell Georgia 24c; NASA 55c; **European Space Agency:** 51cr; NASA 31tr; **Dr Alan Fitzsimmons, Iwan Williams, Donal O'Ceallaigh** 37bcr; **Galaxy Picture Library:** Gordon Garrad 4cl; **Julian Cotton Photolibrary:** Jason Hawkes Aerial Collection 48l; **NASA:** 1, 10tl, 10tr, 10crb, 12c, 16ca, 16b, 18tr, 18bl, 21l, 34ca, 39br, 48cl, 48c, 51tl, 51cla, 52bl, 52c, 52cr, 53br, 53c, 53tl, 54b, 54tr, 55tc, 55bl, 57cl, 57tr, 57cr, 58cl, 58tr, 60ca, 61al, 61c, 61r, 61tr, 61tl, 63c; **Calvin J. Hamilton:** 12c; **COBE Project:** 44bl, 48t; **GSFC:** 58cr; **GSFC/TRACE:** 8cl; **Jet Propulsion Laboratory/Caltech:** 2tl, 3tr, 6 all images, 7 all images,

12bl, 13br, 14t, 14l, 15r, 19rt, 22l, 22c, 23br, 23tr, 23cr, 24b, 24tr, 25b, 25b, 26bl, 26tl, 26c, 27r, 27tr, 28tr, 28bl, 28br, 29tr, 29l, 30c, 30bl, 30cl, 31r, 32br, 32tr, 32bl, 33c, 33tl, 34c, 34tl, 34bl, 35br, 36cb, 36b, 37bl, 48cr, 62bl, 62br, 62bc, 63bl; Calvin J. Hamilton 35tr, 22bl; David Seal 29b, 31bl; Kennedy Space Center 2b; Lockheed Martin 51br; **Johnson Space Flight Center:** 8c, 54c, 54tr; **Kennedy Space Flight Center:** 2b, 18bl, 18c, 20c, 20tl, 63r; **Laboratory for Atmospheres/GSFC:** 58cr; **Lockheed:** 53tr; **Space Sciences Laboratory/Marshall Space Flight Center:** 8t; **Stanford Lockheed Institute/TRACE:** 8bl; **Natural History Musem, London:** 17l; **National Geographic Image Collection:** Don Foley 25cr; **Novosti London:** 50tl, 50bl, 63t; **NSSDC/GSFC/NASA:** 3c, 15tt, 16tr, 17tc, 17tr, 17br, 19cra, 20tr, 21c, 21tl, 39b, 56tl, 57br; Michael Tuttle 20b; **Planet Earth Pictures:** 5tr, 11br, 54r, B. Sidney 10tr, 11bl;

Royal Observatory Edinburgh: 4c, 42tc; David Malin 40bl, 40c, 45bl, 46b; **Science Photo Library:** Agence SPOT 59cl; David Nunuk 5tl, 5c, 39tl; David P. Anderson SMU/NASA 14b; David Parker 60c; Dr Fred Espenak 49bl; Earth Satellite Corp 59cr, 59tr; Francis Gohier 39cr; Frank Zullo 45; Fred Espenak 11c; Jerry Schad 38tl; Michael J. Ledlow 13tr; NASA 50c, 56c, 58b; Rev. Ronald Royer 10b; Siding Spring Observatory, Australia 38c; Simon Fraser/Mauna Loa Observatory 4tr; Smithsonian Institution 43c; Stargazers Radio Telescope 60tr; Tony Hallas 45b; **Courtesy of SOHO/EIT Consortium:** 48cr, 62; ESA/NASA 6bl; 9c; **STScI/AURA/NASA:** 44cra; A. Stern, (Southwest Research Institute); M. Buie, Lowell Observatory, ESA 36c; B. Balick (University of Washington), V. Icke (Leiden University), G. Mellema (Stockholm University) 43tr; Brad Whitmore (STScI) 47br; E. Karkoschka (University of Arizona

32c; Hubble Heritage Team, J. Trauger (JPL) and collaborators 43c; B. Balick, J. Alexander (University of Washington) 43cr; J. Harrington, K. Borkowski (University of Maryland) 43br; J. Hester, P. Scowen, (Arizona State University) 40c; J. Morse (University of Colorado) 42bc; John Clarke, (University of Michigan) 27br; Jon Morse, (University of Colorado), NASA 5br; R. Williams, The HDF Team 49tr; Reta Beebe, (New Mexico State University) 27tr.

Book jacket:
Anglo Australian Observatory: David Malin back l; **NASA/JPL/Caltech:** back flap b, back r, back c; **Royal Observatory Edinburgh:** D. Malin front c; **Science Photo Library:** Ronald Royer spine c; **STScI/AURA/NASA:** Erich Karkoschka (University of Arizona) front flap b; B. Balick, J. Alexander (University of Washington) back cl; J. Hester, P. Scowen (Arizona State University) back cr.